T0338639

INDIGENOUS HEALTH ETHICS

An Appeal to Human Rights

Intercultural Dialogue in Bioethics

ISSN: 2515-7035 (Print)
ISSN: 2515-7043 (Online)

Series Editor: Alireza Bagheri
(Tehran University of Medical Sciences, Iran)

Published

■ Intercultural Dialogue in Bioethics — Vol. 3

Series Editor: Alireza Bagheri

INDIGENOUS HEALTH ETHICS

An Appeal to Human Rights

editors

Deborah Zion
Victoria University, Australia

Linda Briskman
Western Sydney University, Australia

Alireza Bagheri
Lakehead University, Canada

 **World Scientific**

NEW JERSEY · LONDON · SINGAPORE · BEIJING · SHANGHAI · HONG KONG · TAIPEI · CHENNAI · TOKYO

Published by

World Scientific Publishing Europe Ltd.

57 Shelton Street, Covent Garden, London WC2H 9HE

Head office: 5 Toh Tuck Link, Singapore 596224

USA office: 27 Warren Street, Suite 401-402, Hackensack, NJ 07601

Library of Congress Cataloging-in-Publication Data
Names: Zion, Deborah, 1961–   editor. | Briskman, Linda, editor. | Bagheri, Alireza, editor.
Title: Indigenous health ethics : an appeal to human rights / editors,
    Deborah Zion, Linda Briskman, Alireza Bagheri.
Other titles: Intercultural dialogue in bioethics ; v. 3. 2515-7035
Description: New Jersey : World Scientific, [2020] | Series: Intercultural dialogue in bioethics,
    2515-7035 ; vol. 3 | Includes bibliographical references and index.
Identifiers: LCCN 2020020782 | ISBN 9781786348562 (hardcover) |
    ISBN 9781786348579 (ebook for institutions) | ISBN 9781786348586 (ebook for individuals)
Subjects: MESH: Bioethical Issues | Health Services, Indigenous--ethics |
    Human Rights | Health Equity
Classification: LCC R724 | NLM WB 60 | DDC 174.2--dc23
LC record available at https://lccn.loc.gov/2020020782

British Library Cataloguing-in-Publication Data
A catalogue record for this book is available from the British Library.

For any available supplementary material, please visit
https://www.worldscientific.com/worldscibooks/10.1142/Q0253#t=suppl

Printed in Singapore

Acknowledgement & Dedication

Acknowledgement

There are many people who have played important roles in the development of the book. We thank each and every contributor for providing insights from national and international perspectives, for raising issues that beset Indigenous peoples worldwide and discussing the role of bioethics in maintaining or overcoming discrimination, oppression and racism. We thank Joy Quek, editor of World Scientific Publishing, for supporting the book and recognizing its significance to academia and to communities of practice. A special thank you to Dr. Aminath Didi, whose diligence held the project together and brought it to a fine conclusion.

Dedication

We dedicate this book to two outstanding bioethics scholars and advocates for the health of Indigenous peoples. Both died in 2019 and we mourn them deeply.

Barry Smith: Of Māori descent (Te Rarawa and Ngati Kahu), Barry Smith had a stellar career in health analytics and health ethics with the Lakes District Health Board in Rotorua, New Zealand. He worked extensively on prestigious research projects in the areas of ethics, genomic research and smoking cessation. Publications include a book with Martin Tolich entitled *The Politicisation of Ethics Review in New Zealand* (Dunmore Publishing,

2015), the multi-authored *Te Ara Tika: Guidelines for Maori Research Ethics* (HRC, 2010) and *Te Mata Ira: Guidelines for Genomic Research with Maori* (University of Waikato, 2016). Barry was a highly respected member and chair of HRC science assessing committees and sat on the National Heart Foundation research committee. He chaired the Ministry of Social Development Ethics Committee, the Lakes DHB Research and Ethics and Clinical Ethics Committees and was a previous chair of the HRC Ethics Committee. He was also a member of the Advisory Committee on Assisted Reproductive Technology (ACART), the Auckland Regional Tissue Bank Governance Advisory Board and chaired the Royal Society of New Zealand Maori Reference Group that provides advice on Indigenous matters to the Society's project on gene editing. A frequently invited presenter at ethics conferences, he presented at the 2016 WHO Global Ethics Summit in Berlin and the Asia-Pacific Regional Ethics Meeting in Seoul in October 2017. He also contributed to the drafting of National Ethical Standards for Health and Disability Research and to the development of the Te Ara Tika Guidelines for health research conducted with Maori and in the integration of the values described in Te Ara Tika into the Draft Standards.

Leo Pessini: Father Leo Pessini joined the Latin American Episcopal Council-CELAM, in the area of Pastoral Health in 1994 when the II Latin American Meeting was held in Quito, Ecuador. On that occasion, the Latin American Health Pastoral Guide was prepared, which has been updated periodically, under the coordination of Father Adriano Tarrarán, also of the Order of St. Camillus. Later, when the Pastoral Observatory for Latin America-OBSEPAL was created at CELAM, he directed the Bioethics Area since 2007 and contributed with his studies and research to keep the readers of the website of the mentioned Observatory updated. He also collaborated with his publications and with teaching in the training of pastoral agents of health.

Father Leo had a doctorate in Moral Theology, with a specialization in Bioethics. He had specialized in Clinical Pastoral Education and Bioethics at the Center for Bioethics at Edinboro University, Pennsylvania, USA. In Latin America and Brazil, Father Leo played a decisive role in the development and dissemination of bioethics, being the author of numerous works and contributing with many others. In addition, he participated and contributed to bioethical issues in several committees.

Contents

About the Editors

Deborah Zion is an Associate Professor and Chair of the Victoria University Human Research Ethics Committee. She has published widely on the ethics of the treatment of asylum seekers, and on research ethics concerning vulnerable populations, including people with infectious diseases, particularly HIV/AIDS. She currently teaches research ethics, and for many years taught in the area of Global Health.

Linda Briskman holds the Margaret Whitlam Chair of Social Work at Western Sydney University in Australia. She is a human rights researcher and activist in the spheres of Indigenous rights, asylum seeker rights and challenging Islamophobia. She publishes extensively, with recent books including the co-edited *Social Work in the Shadow of the Law* (2018), Federation Press, and *Social Work with Indigenous Communities: A Human Rights Approach* (2014), Federation Press. A current collaborative project is on the wellbeing of young Indigenous people who identify as LGBTIQ.

Alireza Bagheri-chimeh is a physician-clinical ethicist and currently is a research affiliate with the Center for Healthcare Ethics in Lakehead University in Canada and provides ethics consultation at the Regional Health Science Center. He is an Elected Fellow of The Hastings Center (USA) and was an Assistant Professor of medicine and medical ethics in Tehran University of Medical Sciences in Iran. He served as a member of the UNESCO International Bioethics Committee (2010–2018) and on the Board of Directors of the International Association of Bioethics (2009–2014). In 2010, he received the National Razi Medical Research Award in Iran and in 2018, the Bioethics Leadership Award in Canada. He has studied and taught bioethics in Iran, Japan, Belgium, the United States and Canada, which strengthened his understanding of bioethics in a global context. As a palliative care physician, he brings extensive clinical experience to bioethical discussions as well as the promotion of close attention to socio-cultural context, especially in healthcare delivery. As a clinical ethicist working in Indigenous communities in North Western Ontario in Canada, he advocates for equitable access to healthcare without discrimination for Indigenous people.

Contributors

Eselealofa Apinelu is Attorney-General and Indigenous Rights Strategist with the Government of Tuvalu, Funafuti, Tuvalu. She is Founder of Tuvalu Women Infusion for Societal Empowerment (TuWISE) and Education Services Engineers (ESE) Consultancy. She infuses Indigenous and contemporary scientific knowledge to enhance women and youth development with a particular focus on financial literacy and economic self-sufficiency as a means for eliminating poverty and domestic violence in Tuvalu.

Atsushi Asai is Professor of Medicine and Bioethics at the Department of Medical Ethics, Tohoku University Graduate School of Medicine, Japan. He was a visiting research fellow at the Centre for Human Bioethics, Monash University in Australia and also an ethics fellow at the University of California, in San Francisco, USA.

Alireza Bagheri is a physician-clinical ethicist and currently is a research affiliate with the Center for Healthcare Ethics at Lakehead University in Canada. Dr. Bagheri is an Elected Fellow of The Hastings Center (USA). He served as the Vice-chair of the UNESCO International Bioethics Committee (2010–2014).

Sareh Bagherichimeh is a graduate (BSc with Hons) in Genome Biology and Fundamental Genetics and is a research assistant at the Institute of Medicine, University of Toronto, Canada. She is interested in fairness in resource allocation and healthcare delivery to Indigenous people without social discrimination.

Linda Briskman holds the Margaret Whitlam Chair of Social Work at Western Sydney University in Australia. She is a human rights researcher and activist in the spheres of Indigenous rights, asylum seeker rights and challenging Islamophobia. She publishes extensively, with recent books including the co-edited *Social Work in the Shadow of the Law* (2018), Federation Press, and *Social Work with Indigenous Communities: A Human Rights Approach* (2014), Federation Press. A current collaborative research project is Rainbow Knowledge on the wellbeing of young Indigenous people who identify as LGBTQI.

Julia Dammann graduated in Social Anthropology (Magistra Artium), focusing on the potential of incorporating anthropological fieldwork to the field of development cooperation. Her main interest lies within Indigenous cultures and the work of NGOs. As a SASI staff member she has been responsible for coordinating research involving the San communities.

Yasuhiro Kadooka is Professor, Department of Bioethics, Kumamoto University Faculty of Life Sciences, Kumamoto, Kumamoto, Japan. His backgrounds are bioethics and clinical medicine. Currently, he is engaged in research and education on biomedical ethics, ethical review of biomedical research, and clinical ethics support at medical settings.

Etivina Lovo is a Research Fellow (Bioethics and Professionalism) at the Fiji Institute of Pacific Health Research, College of Medicine, Fiji National University. Her work involves research bioethics and clinical ethics education and in research she looks at engaging Pacific Islands Indigenous and cultural principles in research governance.

Darryl Macer is President of the American University of Sovereign Nations, San Carlos, Arizona, USA. In 1990, he founded the Eubios Ethics Institute in Christchurch, New Zealand and in Tsukuba Science City, Japan. In 2005 Bangkok, Thailand was added to the network. The Eubios Ethics Institute is a non-profit group that aims to stimulate the international discussion of ethical issues, and how technology is used in ways consistent with "good life"(eu-bios). It aims at an integrated and cross-cultural approach to bioethics and has a global network of partners.

Richard Matthews is Associate Professor and Theme Lead, Health Advocate and Professional at Bond University. An experienced bioethicist, social justice advocate and philosopher, he researches violence, health ethics and allyship. He is the author of *The Absolute Violation: Why Torture Must be Prohibited* (Montreal & Kingston, MQUP, 2008) along with a number of articles.

Taketoshi Okita is Associate Professor, Department of Medical Ethics, Tohoku University Graduate School of Medicine, Sendai, Miyagi, Japan. His academic backgrounds are bioethics, ethics and philosophy. His focused research fields are matters of public health, especially HIV/AIDS.

Danielle Pacia received her Bachelor of Arts in Bioethics from the University of Alabama and her Master of Bioethics from Harvard Medical School. She was also a 2018 student at the Yale Summer Bioethics Institute as well as an undergraduate recipient of the Hastings Center Emily Murray Fellowship. Her academic interest areas broadly include health policy, genomic ethics, and reproductive justice.

Leo Pessini in memoriam (1955–2019). Leo Pessini was General Moderator of the Camillianum International Institute of Pastoral Health Theology, connected to the Pontifical Lateran University (Rome, Italy). He was also Professor of Bioethics at the Vale do Sapucaí University (UNIVAS), Pouso Alegre, MG, Brazil.

Crystal Pirie is a Nishnawbe kwe from Netmizaaggamig Nishnaabeg and is the Senior Director of Indigenous Collaboration with the Thunder Bay Regional Health Sciences Centre, Ontario, Canada. In this role, her primary focus is advocacy and engagement efforts with the Indigenous communities, as well as federal and provincial authorities to improve healthcare delivery and equity outcomes among Indigenous patients.

Anor Sganzerla is Adjunct Professor of Philosophy at the Postgraduate Program in Bioethics, at the Pontifical Catholic University of Paraná (PUCPR, Curitiba). Coordinator of the International Doctorate in Humanities, visiting professor at the Catholic University of Mozambique (UCM), member of the Center for Technical Studies and the Center for Studies in Bioethics. He conducts research on Hans Jonas and Van Rensselaer Potter.

Barry Smith (1947–2019) descended from Te Rarawa and Ngati Kahu, and had a PhD in sociology, other university qualifications in mathematics and music and a background in university teaching and social research. He was the chairman of the Rotorua Civic Arts Trust, consultant in social research and health research ethics, acting chairman of the Wellington-based Health and Disability Multi-Region Ethics Committee and chairman of the Lakes District Health Board Research and Ethics Committee. He chaired the Royal Society of New Zealand Maori Reference Group that provides advice on Indigenous matters to the Society's project on gene editing and contributed to the re-writing of the New Zealand National Ethics Guidelines. He also worked with the Australasian Human Research Ethics Consultancy.

Masashi Tanaka MD, Department of Medical Ethics, Tohoku University, Graduate School of Medicine, Sendai, Miyagi, Japan. He works as a physician at the department of general internal medicine. His research focuses on matters of clinical ethics, especially decision making.

Izak Van Zyl is a transdisciplinary researcher working across multiple fields, and primarily anthropology, communication science and applied design. Izak's research is stimulated by complex and ethical problems, such as the decolonial project in marginalized communities in Southern Africa. Izak has published a variety of social scientific papers, proceedings and book chapters, and is recognized as a promising young researcher by the South African National Research Foundation.

Diego Carlos Zanella is a specialist in bioethics at Facultad Latinoamericana de Ciencias Sociales (FLACSO, Argentina), with a period of study at Tübingen University (Germany). He is Adjunct Professor of Philosophy at the Postgraduate Program in Humanities at Universidade Franciscana (UFN, Santa Maria), member of the Human Research Ethics Committee and the Animal Use Ethics Committee, member of the Brazilian Society of Bioethics and the current president of Rio-Grandense Bioethics Society. He conducts research on Humanities and Bioethics, especially Van Rensselaer Potter.

Deborah Zion is an Associate Professor and chairs the Human Research Ethics Committee at Victoria University in Melbourne. She has spent many years teaching, investigating and writing about ethics and vulnerable populations, including persons with HIV/AIDS and asylum seekers. She has published extensively in both areas across a range of bioethics, medical and public health journals, looking at questions related to ethics and justice.

Preface
Indigenous Healthcare: A Broken Promise

Deborah Zion,* Linda Briskman† & Alireza Bagheri‡

The legacy of colonialism remains present in the lives of Indigenous peoples around the world. Racism and discrimination continues, despite existing international human rights instruments that focus on the well-being of Indigenous peoples, signaling our failure to eliminate injustice in Indigenous healthcare.

It has been estimated by the United Nations that the world's 370 million Indigenous peoples, residing in approximately 70 countries, are among the world's most marginalized. They are often included in estimates of the poorest peoples, with the poverty gap between Indigenous and non-Indigenous groups increasing in many countries. This influences Indigenous peoples' quality of life and their right to health [UN Department of Economic and Social Affairs 2015]. The position of the

*Associate Professor and Chair Human Research Ethics Committee, Victoria University, Australia.
†Margaret Whitlam Chair of Social Work, Western Sydney University, Australia.
‡Research Affiliate, Center for Healthcare Ethics, Lakehead University, Canada.

xvii

United Nations Permanent Forum on Indigenous Issues is a useful starting point for understanding the global significance of Indigenous peoples.

> Practicing unique traditions, they retain social, cultural, economic and political characteristics that are distinct from those of the dominant societies in which they live. Spread across the world from the Arctic to the South Pacific, they are the descendants — according to a common definition — of those who inhabited a country or a geographical region at the time when people of different cultures or ethnic origins arrived. The new arrivals later became dominant through conquest, occupation, settlement or other means... most other Indigenous peoples have retained distinct characteristics which are clearly different from those of other segments of the national populations [United Nations Permanent Forum on Indigenous Issues 2015].

Based on the diversity of Indigenous peoples, the Permanent Forum states that no official definition of the term Indigenous has been adopted. Rather the UN system has developed understandings of the term based on self-identification by individuals that is accepted by the community; historical continuity with pre-colonial and/or pre-settler societies; strong link to territories and surrounding resources; distinct social, economic or political systems; distinctive language, cultures and beliefs; formation of non-dominant societal groups; and resolve to maintain and reproduce ancestral environments and systems as distinctive peoples and communities [UN Permanent Forum on Indigenous Issues 2015].

The health status of Indigenous peoples worldwide is unacceptable. Through the combination of physical invasion, the importation of diseases, and Indigenous depopulation [McCracken & Phillips 2017], the colonization process continues to have tragic consequences. This influences Indigenous peoples' quality of life and their right to health [UN Department of Economic and Social Affairs 2015]. The recent 2020 *Closing the Gap* report concerning Australia's Aboriginal and Torres Strait Islanders reminds us of the deeply entrenched and harmful effects of colonization. The report, released as we finalize this book, reveals that it is still the case that the life expectancy gap is between 7 and 9 years less than for other Australians, and the rate of mortality for children under five is twice as high [Australian Government 2020].

It is not just the health of Indigenous populations that is under threat, but also other economic, social and political rights, which have, in their turn, an impact on health status, such as the occupation of ancestral lands, often with disastrous consequences. In South America, Indigenous land rights, environments and communities have been threatened by wildfires brought about by land clearing for agriculture or mining [The International Working Group For Indigenous Affairs 2019.] More recently the Tohono O'odham sacred burial ground in Arizona was destroyed to make way for Trump's wall between the US and Mexico [Jankowicz 2020].

There are complex relationships between health, cultural practice, human rights and ethics. This volume is based around two interlinked and overlapping themes that explore these issues: *Health and Colonization* and *Indigenous Knowledges and Research.*

The first theme recognizes that colonization fundamentally undermines human rights, both individually and collectively. There is increasing consensus that health inequalities, disproportionate rates of disease, disability, addiction and violence in Indigenous communities around the world are the result of the colonization process [Reading 2018]. As stated by the World Health Organization's (WHO) Commission on Social Determinants of Health, 'Everyone agrees that there is one critical social determinant of health, the effect of colonization' [Mowbray 2007].

What then follows from this? As the World Health Organization [2017] states, 'understanding health as a human right creates a legal obligation on states to ensure access to timely, acceptable, and affordable healthcare of appropriate quality'. However, it is not just healthcare that is at issue here, but rather the conditions that allow good health to flourish [Toebes 1999]. As human rights are indivisible, the right to health must also encompass all other rights. Article 21 of The United Nations Declaration on the Rights of Indigenous Peoples [UNDESA 2007] affirms this. It states that:

Indigenous peoples have the right, without discrimination, to the improvement of their economic and social conditions, including, inter alia, in the areas of education, employment, vocational training and retraining, housing, sanitation, health and social security.

However, even the 'minimum standards for the survival, dignity and well-being of the Indigenous peoples of the world' introduced by the UN Declaration (Article 43), have not yet been fulfilled.

The second theme of this volume, *Indigenous Knowledges and Research* explores the relationship between these ideas, the creations of different forms of Indigenous bioethics, and the predominantly Western model still used to guide research. Do the insights of both Indigenous and non-Indigenous scholars working in this area have the capacity to transform the field?

Research has had a powerful part to play in making political, cultural and material domination possible. Just as history has often left out the stories of the lives of the oppressed, so too, research of all kinds has been used to 'naturalize' colonization. Drawing on the work of Albert Memmi, Marie Battiste identifies four related strategies used to maintain power over Indigenous peoples and links them to the research context. These are:

(1) stressing real or imaginary differences between the racist and the victim; (2) assigning values to these differences, to the advantage of the racist and to the detriment of the victim; (3) trying to make these values absolutes by generalizing from them and claiming that they are final; and (4) using these values to justify any present or possible aggression or privileges. All these strategies have been the staple of Eurocentric research, which has created and maintained the physical and cultural inferiority of Indigenous peoples [Battiste 2008, p.507].

How do these issues correlate with a Western bioethics? Bioethics, although now incorporating ways of thinking that emphasize both justice and inter-connectiveness, has, in the past, been concerned with a Western view of the self as bounded and rational [Beauchamp & Childress 2009]. Autonomy has been privileged as the core concept, both within and outside the clinic.

As King, Henderson and Stein [1999, p. 15] state, research ethics, based upon this model, is preoccupied with the following [see also Smith, 2006]:

- Balancing principles — autonomy, beneficence, justice, informed consent and confidentiality
- Ethical universalism (not moral relativism) — truth (not stories)
- Atomistic focus — small frame, centered on individuals.

However, this is challenged by models from Indigenous communities, that are centered instead upon:

* Consultation, negotiation and mutual understanding
* Respect, recognition and involvement
* Benefits, outcomes and agreement [AIATSIS 2002, Smith 2006, p. 15].

Within this volume, we explore a range of challenges. Context is important and each of the contributing authors provides broad understandings of the peoples they write about so that bioethics is positioned within that realm, focusing on two main human rights questions, as stated earlier; namely colonization, and questions of subjugated Indigenous knowledges through the lens of research with Indigenous peoples. These are important, given colonialist research with Indigenous peoples, which not only ignored their knowledges but commodified them. A comment by Rosemary Clews [1999] in Canada still resonates. She speaks of a concept of knowledge among First Nations peoples where knowledge and wisdom are gifts from the creator to be used with respect and stewardship.

Chapters are written by either Indigenous authors, each with their own unique perspective and context, or by non-Indigenous writers who draw on Indigenous literature and research. While acknowledging that context matters and that the expression 'Indigenous' is contested, especially as it has been co-opted by non-Indigenous policy makers, we choose to utilize this term. We note that the leadership of Indigenous communities in many countries also employ it, and for us it also helps bypass limitations of the definition to include societies that were not colonized in the formal sense of theoretical understandings but have been subject to colonial-like processes.

Although there exists a collection of bioethical writing that is based upon clinical issues, here we have confined ourselves to subjects discussed by the authors rather than taking on the full gamut of concerns that are common to the field such as abortion, euthanasia, in vitro fertilization and organ donation, which we regard as niche topics that also need to be underpinned by overall Indigenous frameworks that result in health injustices.

Similarly, we do not discuss the question of patient-healthcare provider relationships per se, about which there is increasing literature. The collection instead, in keeping with its human rights and social justice stance, focuses on health justice and health rights and the underpinnings that guide our authors are located in colonial heritage and Indigenous knowledges. Thus, in this volume we undertake to reflect upon the impact of colonization on Indigenous populations, and their responses to it. We also hear the voices of Indigenous writers and their reflections on local knowledge in relation to ethical and cultural frameworks. We hope this work will contribute a different way of seeing the field of bioethics itself.

To establish background to the collection, Chapter 1, written by the three editors, analyzes Indigenous health ethics from a human rights perspective, pointing to examples of global health normative paradigms that fall short of integrating Indigenous human rights, particularly collective rights. The chapter contemplates the leadership that the discipline of bioethics might take to work in solidarity with Indigenous peoples in order to overcome disparities and reverse ongoing colonialist approaches.

In the second chapter, by emphasizing the importance of patient narratives in healthcare, alluded to in Chapter 1, Alireza Bagheri, Crystal Pirie and Sareh Bagherichimeh present several Indigenous patient stories from Canada to illustrate how patients' narrative plays an integral part in healthcare and how these narratives can provide a major vehicle for the transmission of knowledge and practice for marginalized Indigenous populations. The narratives illustrate ongoing racism and discrimination and failure to eliminate injustice.

From Brazil, in Chapter 3, Leo Pessini, Anor Sganzerla and Diego Carlos Zanella analyze universal declarations and discuss prospects for bioethics that guide the promotion and protection of Indigenous peoples in Brazil. The authors argue that dominant models of bioethics are not sufficient for the reality of the Latin American continent. The Indigenous peoples of Brazil are diverse, and their histories of human and environmental relationships have been drastically altered by colonization. The chapter intersects with the second theme of this book on the importance of Indigenous knowledges and the interaction between traditional knowledges and modern-day health practices, which may, in the Brazilian context, result in political misunderstandings to suit the position of the dominant society. A case study reveals how tradition and science can work in tandem, given the will.

The following chapter is from a Japanese perspective where Taketoshi Okita, Atsushi Asai, Masashi Tanaka and Yasuhiro Kadooka discuss the impact of discrimination on Japan's Indigenous peoples, giving prominence to the Ainu and using illustrative case studies. They document the long history of invasion and opposition to the Ainu. Discussed is the effect of phrenology on biological and social life, with exploitation of Ainu bodies for research, including removal of the dead for scientific purposes and taking blood from the living. Eugenics as 'sham science' was used to justify discrimination and objectification.

Analysis of the devastating impact of the eugenics movement on Indigenous Peoples is continued in Chapter 5 by Darryl Macer from the United States. Macer presents a compelling paper on the legacies of this colonial movement, contextualized within the San Carlos Apache Nation. Macer's chapter looks at the historical interface between eugenics, race and colonization of Native American peoples. By tracing the history of colonization, imposition and eugenics ideology over time, he leads us into present day problematics.

In Chapter 6 from Tuvalu, Eselealofa Apinelu outlines the conundrum of individual human rights, particularly human rights law, in a society where interdependency is a cultural norm. Tuvalu in the Pacific is an interesting case study as although colonized by the British, unlike settler society nations the Indigenous population remained the majority. Nonetheless, the colonialism that took hold acted against collective traditions by privileging individual rights. Incorporating a legal case in the chapter illustrates the tension between custom and law and the impact of this tension on well-being. With bioethics dominated by colonial Western values and leaning toward individual rights, Apinelu argues that this impacts negatively on the collective and individual health and well-being. Collective interdependency as a positive cultural value suggests the need for prioritizing in law.

From Canada, in Chapter 7, Richard Matthews discusses the implications of colonialism and Indigenous cultural genocide for healthcare ethics, which are currently under-explored. He suggests that colonialism is a process that is ethnocidal by its nature. Located in Northern Ontario, this reflective chapter is likely to have global resonance as questions of cultural genocide through ongoing health colonialism reverberate in many countries. His chapter is highly informative and at the same time provocative in the challenges presented to uncritical bioethicists and critiques of long-accepted norms and principles, such as beneficence that Matthews posits is skewed to the interests of dominant settler-colonialists. Problems with public health ethics are discussed. With the passing of real time since the 'colonial project' in settler societies, this chapter reveals the impacts of colonialism and the entrenchment within today's societies, underpinned by racism.

Writing from New Zealand, in Chapter 8, Barry Smith attributes the social and economic conditions of the Maori to colonization processes, furthered by ongoing assumptions of dominant groups that devalue Maori constructs. In his comparative study across New Zealand, Australia and Canada, Smith introduces the concept of 'Social Equipoise' to ethics oversight in research. He provides evidence of social inequality and poor health and social outcomes noting that in a study of 23 developed countries, New Zealand was ranked fifth from bottom, with Australia, and less so Canada, exhibiting serious degrees of income inequality. Smith's exploration of Indigenous research ethics across three countries draws out existing deficits and prospects.

Turning to the San peoples of Southern Africa in Chapter 9, Julia Dammann, Izak Van Zyl and Danielle Pacia continue the theme of how empirical and social scientific research has been imbued with Western hegemonic values and imperialist constructs. This results in Indigenous communities becoming vulnerable in the face of Western centered research. For the San, exploitation includes resource scarcity, differences in notions of autonomy, cultural sensitivities and language barriers. The authors present a critical overview of bioethical thought and argue for the collective over the individual in ethical decision-making. The San Code of Research Ethics was introduced in 2017 to protect Indigenous rights around research data.

Fijian researcher Etivina Lovo (Chapter 10) also challenges Western constructions of research ethics principles, through the lens of Pacific Island countries where research involving human participants has increased. She outlines tensions of research ethics principles that originated in developed countries and which influenced international guidelines for research involving humans. The inadequate protection of participants is illustrated by the 2002 case involving the company Autogen in Tonga, where a deep divide emerged between Autogen's view of informed consent and contrasting views of the Tongan people. Emerging from her study of Indigenous research frameworks in three Pacific Island nations, Lovo identifies Indigenous principles that are not included in internationally accepted research principles involving humans.

In health policies, like other domains, it is important to examine the frame of laws and regulations aimed at protecting human rights. Although, the United Nations Declaration on the Rights of Indigenous Peoples [UNDESA 2007] along with other international and country-specific efforts have contributed to change, they have not yet been successful in elimination of health inequity, discrimination and injustice to Indigenous peoples. This work seeks to explore reasons for this, and more significantly, the lived experience of Indigenous peoples in relation to the right to health, and through it, all other rights.

References

AIATSIS (2002). *Guidelines for Ethical Research in Australian Indigenous Studies*. Available from: <https://aiatsis.gov.au/research/ethical-research/guidelines-ethical-research-australian-indigenous-studies>.

Australian Government (2020). *Closing the Gap Report, 2020*. Available from: <https://ctgreport.niaa.gov.au/sites/default/files/pdf/closing-the-gap-report-2020.pdf>.

Battiste, M. (2008). Research Ethics for Protecting Indigenous Knowledge and Heritage: Institutional and Researcher Responsibilities. In Denzin, N. K., Lincoln, Y. S. and Smith, L. T. (eds.) *Handbook of Critical and Indigenous Methodologies* (SAGE, Thousand Oaks, California) pp. 497–510.

Beauchamp, T. and Childress, J. (2009). *Principles of Biomedical Ethics* (Oxford University Press, New York).

Clews, R. (1999). Cross-cultural Research in Indigenous Rural Communities: A Canadian Case Study of Ethical Challenges and Dilemmas, *Rural Social Work* 4: pp. 26–33.

Jankowicz, M. (2020). Border Officials are Blowing Up a Sacred Native American Burial Site to Make Way for Trump's Border Wall, *Business Insider Australia*.

Available from: <https://www.businessinsider.com.au/trump-border-wall-native-american-burial-site-arizona-20202?fbclid=IwAR1Mny1rf2thHH3bE wzerN7CWuUmUI3pYII0bGuL_igCSqlIaGbeOVBKCQ4&r=US&IR=T>.

King, N. M., Henderson, G. E. and Stein, J. (1999). Introduction. Relationships in Research. A New Paradigm. In King, N. M., Henderson, G. E. and Stein, J. (eds.) *Beyond Regulation: Ethics in Human Subjects* Research (University of North Carolina Press, North Carolina) pp. 1–20.

McCracken, K. and Phillips, D. (2017). *Global Health: An Introduction to Current and Future Trends*, 2nd edition (Routledge, Abingdon, Oxon, New York).

Mowbray, M. ed. (2007). *Social Determinants and Indigenous Health: The International Experience and its Policy Implications.* (Geneva, Switzerland: World Health Organization Commission on Social Determinants of Health).

Reading, C. (2018). Structural Determinants of Aboriginal Peoples' Health. In Greenwood, M., Leeuw, S. D. and Lindsay, N. M. (eds.) *Determinants of Indigenous Peoples' Health: Beyond the Social* (Canadian Scholars).

Smith, L. T. (2006). Researching in the Margins Issues for Māori Researchers: A Discussion Paper, *AlterNative: An International Journal of Indigenous Peoples* 2(1): pp. 4–27.

Smith, L. T. (2012). *Decolonizing Methodologies, Research and Indigenous Peoples* (Zed Book, London) p. 20.

The International Working Group for Indigenous Affairs (2019). Indigenous People's Rights, Key in Stemming the Amazon Fires. Available from: <https://www.iwgia.org/en/focus/land-rights/3433-indigenous-peoples-rights-key-in-stemming-amazon-fires.html>.

Toebes, B. (1999). Towards an Improved Understanding of the International Human Right to Health, *Human Rights Quarterly* 21: pp. 661–679.

United Nations Department of Economic and Social Affairs (UNDESA) (2007). United Nations Declaration on the Rights of Indigenous Peoples 2007. Available from: <https://www.un.org/development/desa/indigenouspeoples/declaration-on-the-rights-of-indigenous-peoples.html>.

United Nations Department of Economic and Social Affairs (UNDESA) (2015). *State of the World's Indigenous Peoples: Indigenous Peoples' Access to Health Services.* Available from: <https://www.un.org/development/desa/indigenouspeoples/wp-content/uploads/sites/19/2018/03/The-State-of-The-Worlds-Indigenous-Peoples-WEB.pdf>.

United Nations Permanent Forum on Indigenous Issues (2015). *Who are Indigenous Peoples?* Available from: <https://www.un.org/esa/socdev/unpfii/documents/5session_factsheet1.pdf>.

World Health Organization (2017). Human Rights and Health. Available from: <https://www.who.int/news-room/fact-sheets/detail/human-rights-and-health>.

CHAPTER ONE

Indigenous Bioethics and Indigenous Rights

Linda Briskman,* Deborah Zion† & Alireza Bagheri‡

Abstract

Issues related to the health of Indigenous peoples are deeply connected to the ongoing legacy of colonization and to complex issues about access to human rights. In this chapter we explore how the field of bioethics can contribute to solutions through a consideration of Indigenous perspectives on health and the ongoing effects of colonization. We outline critiques of normative approaches to human rights and to health restoration, including resistance to recognition of collective rights in theory and practice, and the subjugation of Indigenous knowledges by dominant cultures.

Introduction

Attention to the grave state of Indigenous health, particularly in settler colonial societies, is subject to increasing global attention from academics, policy makers, and health practitioners. Most significantly, these

---

* Margaret Whitlam Chair of Social Work, Western Sydney University, Australia.
† Associate Professor and Chair, Human Research Ethics Committee, Victoria University, Australia.
‡ Research Affiliate, Center for Healthcare Ethics, Lakehead University, Canada.

issues are the focus of Indigenous peoples throughout the world who work collaboratively to challenge paradigms that have worked against their interests.

The World Health Organization [2017] emphasizes the interconnection of health and human rights. Understanding health as a human right creates a legal obligation on states to ensure access to timely, acceptable, and affordable healthcare of appropriate quality. Despite the attention to ongoing disadvantage, there has been slow progress in attaining health justice for the world's Indigenous peoples. Solutions remain elusive, in part resulting from contested approaches to 'progress'. Another barrier is that despite Indigenous peoples' more holistic understandings of health, there are sharp divisions among those who see health related to spirituality and connection to land — partially seen to be redressed by acknowledging the impact of colonialism and through de-colonizing measures such as privileging Indigenous knowledges — as opposed to remedies proposed by dominant constructions of health. With increasing trends toward neo-liberalism in Western industrialized societies, the emphasis on cultural diversity and pluralism becomes minimized in the political sphere and in policy deliberations and practice. The neo-liberal turn results in deprecation of the concept of Indigenous self-determination and Indigenous governance and contributes to media discourses that blame Indigenous people for their predicament [Briskman 2014]. Regrettably, existing international documents such as the Universal Declaration of Human Rights (1948), and the UN Declaration on the Rights of Indigenous People [United Nations 2007], which introduces 'minimum standards for the survival, dignity and well-being of the Indigenous peoples of the world' (Article 43), have not been successful in elimination of health inequity, discrimination and injustice.

The importance of a holistic approach was emphasized by the UN Permanent Forum on Indigenous Issues at its Eighteenth Session of 2019:

> Indigenous knowledge systems contribute directly to sustaining biological and cultural diversity, poverty eradication, conflict resolution, food security and ecosystem health and serve as the foundation of Indigenous peoples' resilience to the impact of climate change [United Nations Permanent Forum 2019].

The intersection of bioethics and the human right to health provides a promising framework for seeking answers. By 'health' we mean both healthcare, and the conditions that promote and ensure human flourishing. In this context, the interaction between the global and the local is of paramount importance. While it is possible to make some over-arching connection between rights, colonialism and bioethics, it is the context in which these issues take place that enhances understandings and promotes local solutions.

To set the scene for the chapter, we offer the following propositions about Indigenous health: it is a human rights issue; it is located in the politics of colonization, overtly past and implicitly present; and that subjugated Indigenous knowledges require privileging over paradigms of dominant state actors. As non-Indigenous academics, we share with our Indigenous friends and colleagues deep concerns about the state of Indigenous health and the prevailing approaches to its resolution. We take as our starting point that non-Indigenous people need to be part of the solidarity movement as a way of acknowledging our own privileged positions as well as sharing responsibility. Although we were not directly involved in colonizing processes, we reject calls by the 'new right' which refuses to take responsibility for past wrongs and current injustices. Rather we assert that societies that have benefited from the dispossession of Indigenous peoples are obliged to take some responsibility in partnership with Indigenous populations.

Human Rights Discourse and Indigenous Peoples

Put together, human rights instruments, global commitments and even large-scale endeavors by nation states have failed to effectively deliver change. As Henry Shue [1995, p. 15] states, 'a proclamation of a right is not the fulfilment of a right any more than an airplane schedule is a flight'. Dahre [2010] posits that inadequacies of human rights for Indigenous peoples results from a conceptual blindness — a gap between normative principles and social practice. He argues that the human rights vision is a Western liberal notion, albeit presented as neutral and universal. Thus, human rights do not represent an undisputable universal moral value independent of context; they are at their core about politics. Criticisms of human rights, particularly legalistic and Western dominated, leads further into the contested realm of universalism versus relativism. Under the cultural relativism discourse, human rights ought to incorporate diverse cultural understandings. Although it is beyond the scope of the current project to delve fully into this debate, we note that it is important to have a *critical* human rights approach that moves away from, yet can draw upon, legalistic universal constructs.

Dahre [2010] further debunks the notion that universal human rights are available for everyone, noting that the social and political reality for Indigenous peoples tells a different story where discrimination and marginalization are still part of their lives. We concur with his argument that the discourse of human rights is not necessarily inadequate or meaningless, but that we are required to be aware of limitations in problem-solving, including questions of discrimination and oppression. Further, the United Nations Human Rights Council [2016, p. 3] adds that 'forced assimilation; political and economic marginalization; discrimination and prejudice; poverty; and other legacies of colonialism have also led to a lack of control over individual and collective health'.

The issue of rights-negation in the Indigenous realm is compounded when advocating for collective rights, also known as third-generation rights. Unlike first- and second-generation rights, collective rights are not enshrined in any UN Covenant. They are rights that are defined at a collective level such as belonging to a community, population, society or nation [Ife 2012]. It is on the basis of belonging that certain rights attach, both to individuals within such collectives, and to the collectives themselves. They are particularly relevant to Indigenous societies in the quest for self-determination and as a challenge to colonial practices. They garner little support in mainstream Western societies, which privileges civil and political (first-generation) rights, and to a lesser degree economic, social and cultural (second-generation) rights. Collective rights are particularly significant in emphasizing contexts in which human rights have meaning. However, it is not necessarily the case that this is a conflict between individual rights and group rights. In particular, health in its holistic sense is frequently viewed by Indigenous peoples as both an individual and collective right that is determined by community, land and the natural environment [UN Human Rights Council 2016].

Contradictions exist in discourses about human rights. Resistance to collective rights is part-illustrative of the problems of legal discourse. Such interpretations tend to be 'top-down' and rely on human rights declarations, conventions, charters and laws, which are inevitably drafted by powerful people such as politicians and lawyers; this results in an elite telling others what their rights are [Briskman & Ife 2018]. Nonetheless, the World Health Organization [2017] states that a human rights approach to health provides clear principles for setting and evaluating health policy and service delivery, which targets discriminatory practices and unjust power relations at the heart of inequitable health outcomes.

Furthermore, an emphasis on a decontextualized view of autonomy in certain strands of human rights discourse negates the relationship between collective and cultural rights and individual flourishing. According to Will Kymlicka [1995] cultural rights in fact defend the autonomy of individuals. They do so by enabling persons within such groups to make meaningful decisions about their life plans. Kymlicka [1991, p. 197] states that cultural membership 'allows for meaningful individual choice'. Thus, a person is not free unless he or she has a range of options to choose from that matter to them, and the importance of cultural context is that such persons can relate the social roles they choose to their own lives in a significant way. [Kymlicka 1995, pp. 26–35].

It is not just the issue of collectivity that is of significance when discussing Indigenous rights and ethics, but the way in which these ideas are expressed. The importance of narrative, in which ethical questions and issues are embedded in stories reflects a widely held view that Indigenous peoples see health, mental health and well-being in a more holistic way than Western societies [Phelan 2013], a stance expanded in Chapter Two, 'Narrative Ethics in Indigenous Health.' It is worth emphasizing that the Permanent Forum on Indigenous Issues states that the right to health 'materializes through the well-being of an individual as well as the social, emotional, spiritual and cultural well-being of the whole community' [UN Human Rights Council 2016]. With more 'scientific' notions of health dominating, Indigenous peoples' conceptualizations are frequently marginalized and subjugated, resulting in poor outcomes that do not take into account complex historical contexts.

Universal propositions on the health of all peoples are foundational to the 1948 Universal Declaration of Human Rights (UDHR). Article 25(1) states:

> Everyone has the right to a standard of living adequate for the health and well-being of himself and of his family, including food, clothing, housing and medical care and necessary social services, and the right to security in the event of unemployment, sickness, disability, widowhood, old age or other lack of livelihood in circumstances beyond his control.

More than seventy years after the enactment of the UDHR, these propositions and undertakings have not been achieved for Indigenous peoples throughout the world, resulting in solidarity, resistance and activism from local, national and global Indigenous collectives. As Ife [2012, p. 16] suggests, despite the UDHR being impressive and inspirational, 'it is not holy writ and it can and should be subject to challenge in different times, as different voices are heard, and different issues are given priority'.

The Problematics of Dominant Solutions

There have been various attempts to solve health inequities on a global scale and the question arises as to why there has to date been so little success for the health status of Indigenous peoples and recognition of their approaches to the right to health.

Considerable global attention focused on the Millennium Development Goals and their subsequent replacement by the Sustainable Development Goals. In this chapter however, we avoid the development paradigm, which implies that 'inferior' cultures can advance to the level of modern industrialized nations with the aid of both technology and international aid. Reinforcing this, the proliferation of non-government organizations (NGOs) in communities deemed less developed, have failed to deliver parallel health status to Indigenous communities.

Now at the end of their tenure (2000–2015), the MDGs have failed to deliver on many of the decade-long promises with ambitious targets of reducing child mortality, improving mental health, combatting HIV/AIDs, malaria and other diseases. Despite being more grounded in content than many international UN instruments, they were, however, dominated by international organizations and technocratic approaches to change. For Indigenous peoples there were a number of emphases, including realization of individual and collective rights and access to services, including maternal and child healthcare. Yet the MDGs were rights-limited in falling short of recognizing the rights of Indigenous peoples at all levels and with scant attention to collective rights such as to land and natural resources. The lofty ideals of the MDGs were perhaps little more than beacons of hope or what Hulme [2007, p. 2] describes as 'the world's biggest promise, a global agreement to reduce poverty and human deprivation at historically unprecedented rates through collaborative multilateral action'. Or as Saith [2006] suggests, the MDGs envelop us in a cloud of soft words, good intentions and moral comfort, giving well-meaning persons a sense of solidarity and purpose. At the end of the process, it became clear that health disparities between Indigenous and non-Indigenous populations continued.

The Sustainable Development Goals (SDGs), which superseded the MDGs and are operational at the time of writing, were introduced in 2015 and potentially have more relevance to Indigenous well-being. The SDGs are a set of 17 Global Goals that are measured by progress against 169 targets covering social issues such as poverty, food security, health, education, climate change, gender equality, and social justice. Goal 3 promulgates healthy lives and the promotion of well-being for all ages, as well as calling on states to work toward achieving universal health coverage. Although all 17 goals are relevant for Indigenous peoples, only four out of 230 indicators specifically mention them [Cultural Survival 2017]. As a global call to action with an end-date of 2030, the success of the SDGs for all of humankind has a long implementation journey ahead before their successes or lack thereof can be evaluated.

Work on the Social Determinants of Health is a well-established undertaking by many public health scholars. In 2005, the World Health Organization established the Commission on Social Determinants of Health to address what were seen as avoidable health inequalities related to the social causes of poor health. In 2008 the Commission made three broad recommendations for action: improving daily living conditions; tackling inequitable distributions of power, money and resources; and understanding the problem and assessing the impact of action [McCracken

& Phillips 2017]. Part of this mission involves identifying problematics and progress from data sets. Yet, collecting national and international data on Indigenous health is contentious. Although data is available from some countries, there is little consistency in format and in many parts of the world reliable national-level data is non-existent or not considered reliable [McNeish & Eversole 2005]. Although the social determinants model garners significant levels of support, it fails to address structural changes necessary for adoption that would require government support that has fallen short of active commitment [McCracken & Phillips 2017].

An international network that aims to improve the collection, analysis and use of health information for Indigenous populations has endeavoured to provide some consistency on Indigenous health measurement but at this stage only includes Australia, Canada, New Zealand and the United States [Australian Institute of Health and Welfare 2018].

As noted by the World Health Organization [2017], disadvantage and marginalization serve to exclude certain populations in societies from enjoying good health. It notes that a focus on disadvantage reveals evidence of those who are exposed to greater rates of ill-health and face significant obstacles to accessing quality and affordable healthcare; this includes Indigenous populations. It further states that these populations may be subject to laws and policies that compound their marginalization and make it harder for them to access healthcare prevention, treatment, rehabilitation and care services.

Challenges to Bioethics: Reorienting the Field

Through the emphasis on the right to health, we seek to redress a serious imbalance in the field of bioethics, sometimes referred to as 'the normativity of whiteness', based as it is in an historical preoccupation with what Steven Miles [2003] has referred to as 'heroic individualism'. In the same vein, Catherine Myser [2003, p. 2] states:

> By not seeing or locating the whiteness in bioethics, by theorising from this unself-reflecting white standpoint, and by extending its cultural capital into bioethics policies and practices, we risk functioning as cultural colonisers who do violence to social justice concerns related to race and class.

The standpoint of 'whiteness' has undergone a robust challenge in the last twenty years, particularly with the emergence of public health ethics, 'primarily being focused not on liberty but on a broad account of social justice and health equity in particular' [Wild & Dawson 2018]. This approach thus expands considerations of medical ethics — and the ethics of the western clinic in particular — as the core of bioethics itself [Dawson 2010]. Such a formulation does not take into account the meso, or macro issues related to state-based institutions, global politics, and the history of colonization, and therefore is constricted to philosophical reasoning that is decontextualized and runs the risk of being irrelevant to much of the world's population [Dawson 2010].

To elaborate on how consideration of Indigenous bioethics might further enrich this debate, Stevenson and Murray suggest that it is not only the increasing focus on public health and upon justice as central to the discipline, but also through the use of narrative rather than abstract principles as a method through which to engage with bioethical issues. They state that:

> A story... calls for a hermeneutic act, an interpretation...before this is possible, it must presume a speech act, a narration, in which and by which the story is both spoken and heard. What begins to surface...is the pressing question of how to speak ethically, and how to address, for example, the disproportionate burden of illness among Aboriginal populations, or how to provide ethically appropriate healthcare [Stevenson & Murray 2016, p. 53].

What challenges then do a consideration of Indigenous issues bring to bioethics? The study of bioethics is in itself multi-faceted. It has been referred to as a field, rather than a discipline. As Sarah Chan [2015, p. 18] states:

> Bioethicists sometimes see themselves as divided up into schools of thought characterized by a particular method or approach. From a more general perspective, however, we might argue that the most significant divide in bioethics is not between virtue ethics and consequentialism, or continental and Anglo-American philosophy, but between theory and practice: bioethics-as-philosophy versus bioethics-as-policy.

Reversing Colonialism to Privilege Indigenous Knowledges and Restore Human Rights

A postcolonial analysis is important in order to understand how colonization and racialization create and sustain health inequities faced by Indigenous peoples [Hojjati *et al.* 2018]. It is also significant when discussing Indigenous knowledges for advancement of the health of Indigenous peoples. There is a way to go in influencing the next generation of health and welfare practitioners and academics, whom, although perhaps more informed than previous generations, are yet to have sufficient guidance for incorporating Indigenous knowledge in tertiary education health programs and in health and welfare practice. One cutting edge endeavor to do so is in the work of Zubrzycki *et al.* [2014] who have developed a 'road map' for the development and delivery of Aboriginal and Torres Strait Islander ways of knowledge, being and doing in Australian Social Work curricula.

The issue of decolonization also profoundly effects what we consider ethical practice to be, and what Benatar, Gill and Bakker [2009] refer to as the micro, meso and macro levels of research. By this the authors are referring to the levels of individual interaction, the institutions that support research, and are in turn being researched, and large scale social processes such as colonization, migration and climate change, and it is on all of these levels that the process of decolonization of research needs to occur. Similarly, Ferdinand *et al.* [2018] draw upon the work of Guillemin and Gillam and suggest that researchers working in this field must draw upon reflexivity to achieve decolonization. They develop a taxonomy with four levels, stating:

> The first (type) is self-reflexivity, or how the individual recognises their own biases, assumptions, and ways of working. This reflexivity is tied to the second type, which is interpersonal reflexivity or the ways in which the research works with or collaborates with others and incorporates self-awareness and building trust and rapport. The third type is collective reflexivity, which examines participation in research and the relative roles of the researcher and the community. Finally, institutional reflexivity must also be considered as necessary to facilitate embedded changes in academic research practices such as funding allocation, organizational

partnerships, and patterns of knowledge dissemination [Ferdinand *et al.* 2018, p. 166].

These ways of working speak to the foundational work of New Zealand Indigenous scholar, Linda Tuhiwai Smith, who states that research itself is something of a dirty word within the Indigenous context [Smith 2012]. She outlines the way in which research can continue the work of colonization by privileging Western paradigms of thought rather than beginning with Indigenous perceptions of a life world, and perceptions of what constitutes a problem within such collectivities. The field, she suggests, must seek to recover and validate Indigenous methods of enquiry and ways of knowing.

Finding a way forward to restore the right to health of Indigenous peoples involves reflecting and acting on the historical and ongoing effects of colonization, as well as reintegrating Indigenous reflections in relation to cultural and ethical frameworks. By doing so, there is also the potential to transform the intellectual traditions of bioethics and human rights scholarship.

Indigenous Activism and UN Declaration on the Rights of Indigenous Peoples

Within nation states and as global collectives, Indigenous peoples have formed organizations to fight for their rights and for justice across many spheres of being, including land justice, health justice, education and employment rights, and overcoming disparities in the criminal justice sphere. In each of these fields, health status is influenced by the intersection of disadvantage and organizations run by and for Indigenous peoples have expressed solidarity and focused on fight-back against repressive policies. According to Havemann [1999, p. 470], a politics of embarrassment has characterized Indigenous endeavours to gain rights.

Our chapter ends by referring to the UN Declaration on the Rights of Indigenous Peoples, a declaration that had such a bumpy road to acceptance due to resistance by Western nations that its aspirations were almost not recognized. The Declaration followed the establishment in 1982 of the UN Working Group on Indigenous Populations, with one of its key goals to formulate a Declaration. This resulted in Indigenous peoples from around the world visiting Geneva annually to articulate their claims [Anaya 2007]. The Declaration offered promise as it moved from dominant approaches to health to ones that were more Indigenous-centered and premised on rights. The adoption of the 2007 UN Declaration on the Rights of Indigenous People was a victory for Indigenous peoples and followed two decades of negotiation and opposition from a number of Western countries, including the United States, Canada, Australia and New Zealand. Despite the strength of the health tenets in the Declaration, many are far from realized. For example, Article 24 states:

1. Indigenous peoples have the right to their traditional medicines and to maintain their health practices, including the conservation of their vital medicinal plants, animals and minerals. Indigenous individuals also have the right to access, without any discrimination, to all social and health services.
2. Indigenous individuals have an equal right to the enjoyment of the highest attainable standard of physical and mental health. States shall take the necessary steps with a view to achieving progressively the full realization of this right [United Nations 2007].

Conclusion

It is clear that change is slow, and that there is still much to be achieved. The Declaration does not stand alone in remaining unfulfilled. It sits alongside other United Nations enshrinements such as the 1966 International Covenant on Economic, Social and Cultural Rights, which outlines the social and economic conditions required to live a life of dignity and freedom including health, food, water and a healthy environment. Another example yet to be realized is Article 3(3) of the 1990 Convention on the Rights of the Child that explicitly calls on state parties to the Convention to ensure that 'the institutions, services and facilities responsible for the care or protection of children shall conform with the standards established by competent authorities, particularly in the areas of safety, health…'.

Returning to the field of bioethics, the UNESCO Declaration on Bioethics and Human Rights [2005] recognizes that health does not solely depend on scientific and technological developments but takes into account psychosocial and cultural factors. It further speaks of the desirability of developing new approaches to social responsibility to ensure that progress in science and technology contributes to justice, equity and the interests of humanity. The UNESCO Declaration emphasizes respect for human dignity and human rights and states that 'human dignity, human rights and fundamental freedoms are to be fully respected' (Article 3).

Finally, we echo the words of the UN Human Rights Council that calls out racism. The Council states:

> States often turn a blind eye to racism in healthcare settings, even in the presence of pervasive, persisting evidence that Indigenous peoples are treated discriminatorily. States should take measures to ensure equal access to treatment and healthcare facilities within their jurisdiction, as well as to protect Indigenous peoples from discrimination perpetrated by third party healthcare providers [2016, p. 9].

References

Anaya, S. J. (2007). The UN Declaration on the Rights of Indigenous Peoples: Towards Re-empowerment, *Jurist*. Available from: <https://www.jurist.org/commentary/2007/10/un-declaration-on-rights-of-indigenous-2/>.

Australian Institute of Health and Welfare (2018). International Group on Indigenous Health Measurement. Available from: <https://www.aihw.gov.au/our-services/international-collaboration/international-group-indigenous-health-measurement>.

Benatar, S. R., Gill, S. and Bakker, I. C. (2009). Making Progress in Global Health: The Need for New Paradigms, *International Affairs* 85(2): pp. 347–372.

Briskman, L. (2014). *Social Work with Indigenous Peoples: A Human Rights Approach* (The Federation Press, Sydney).

Briskman, L. and Ife, J. (2018). Extending Beyond the Legal: Social Work and Human Rights. In Rice, S., Day, A. and Briskman, L. (eds.) *Social Work in the Shadow of the Law*, 5th edition (The Federation Press, Sydney) pp. 244–276.

Chan, S. (2015). A Bioethics for All Seasons, *Journal of Medical Ethics* 41(1): pp. 17–21.

Cultural Survival (2017). What do the Sustainable Development Goals Mean for Indigenous Peoples, 18 July. Available from: <https://www.culturalsurvival.

org/news/what-do-sustainable-development-goals-mean-indigenous-peoples>.

Dahre, U. J. (2010). There are No Such Things as Universal Human Rights — On the Predicament of Indigenous Peoples, for example, *The International Journal of Human Rights* 14(5): pp. 641–657.

Dawson, A. (2010). The Future of Bioethics. Three Dogmas and a Cup of Hemlock, *Bioethics* 24(5): pp. 218–225.

Ferdinand, A., Oyarce, A., Kelaher, M. and Anderson, I. (2018). Reflections on Ethics in Indigenous Health Research in Chile, *Revista Latinoamericana de Bioética* 18: pp. 162–184, DOI: 10.18359/rlbi.3451.

Havemann, P. (1999). Indigenous Peoples, the State and the Challenge of Differentiated Citizenship: A Formative Conclusion. In Havemann, P. (ed.) *Indigenous Peoples' Rights in Australia, Canada and New Zealand* (Oxford University Press, Auckland) pp. 468–475.

Hojjati, A., Beavis, A. S. W., Kassam, A., Choudhury, D., Fraser, M., Masching, R. and Nixon, S. A. (2018). Educational Content Related to Postcolonialism and Indigenous Health Inequities Recommended for All Rehabilitation Students in Canada: A Qualitative Study, *Disability and Rehabilitation* 40(26): pp. 3206–3216, DOI: 10.1080/09638288.2017.1381185.

Hulme, D. (2007). *The Making of the Millennium Development Goals: Human Development Meets Results Based Management in an Imperfect World*, Brooks World Poverty Institute, BWPI Working Paper, December 16. Available from: <https://sustainabledevelopment.un.org/content/documents/773bwpi-wp-1607.pdf>.

Ife, J. (2012). *Human Rights and Social Work: Toward Rights-Based Practice* (Cambridge University Press, Cambridge).

Kymlicka, W. (1991). *Liberalism, Community and Culture* (Clarendon Press, Oxford).

Kymlicka, W. (1995). *Multicultural Citizenship. A Liberal Theory of Minority Rights* (Clarendon Press, Oxford).

McCracken, K. and Phillips, D. R. (2017). *Global Health: An Introduction to Current and Future Trends* (Routledge, Abingdon).

McNeish, J. A. and Eversole, R. (2005). Introduction: Indigenous Peoples and Poverty. In Eversole, R., McNeish, J. A. and Cimadamore, A. D. (eds.) *Indigenous Peoples and Poverty: An International Perspective* (Zed Books, London).

Miles, S. (2003). Playing in the Dark: Whiteness and the Bioethics Imagination, *The American Journal of Bioethics* 3(2): p. 12.

Myser, C. (2003). Differences from Somewhere: The Normativity of Whiteness in Bioethics in the United States, *The American Journal of Bioethics* 3(2): pp. 1–11.

Phelan, J. (2013). Narrative Ethics. In *The Living Handbook of Narratology*. Available from: <http://www.lhn.uni-hamburg.de/article/narrative-ethics# Lothe>.

Saith, A. (2006). From Universal Values to Millennium Development Goals: Lost in Translation, *Development and Change* 37(6): pp. 1167–1199.

Shue, H. (1995). *Basic Rights: Subsistence, Affluence, and U.S. Foreign Policy* (Princeton University Press, Princeton).

Smith, L. T. (2012). *Decolonizing Methodologies: Research and Indigenous Peoples* (Zed Books, London).

Stevenson, S. A. & Murray, S. J. (2016). Aboriginal Bioethics as Critical Bioethics: The Virtue of Narrative, *The American Journal of Bioethics* 16(5): pp. 52–54.

UNESCO (2005). Universal Declaration on Bioethics and Human Rights. Available from: <http://portal.unesco.org/en/ev.php-URL_ID=31058&URL_DO=DO_TOPIC&URL_SECTION=201.html>.

United Nations (2007). UN Declaration on the Rights of Indigenous Peoples. Available from: <https://www.un.org/development/desa/indigenouspeoples/wp-content/uploads/sites/19/2018/11/UNDRIP_E_web.pdf>.

United Nations Human Rights Council (2016). *The Right to Health and Indigenous Peoples, with a Focus on Children and Youth*. Draft Study of the Expert Mechanism on the Rights of Indigenous Peoples, July 7. Available from: <https://www.ohchr.org/EN/Issues/IPeoples/EMRIP/Pages/Health Study.aspx>.

United Nations Permanent Forum on Indigenous Issues (2019). *Indigenous Peoples, Indigenous Voices*, Fact Sheet. Available from: <https://www.un.org/esa/socdev/unpfii/documents/5session_factsheet1.pdf>.

Wild, V. and Dawson, A. (2018). Migration: A Core Public Health Ethics Issue, *Public Health* 158: pp. 66–70.

World Health Organization (2017). *Human Rights and Health*, December 29. Available from: <https://www.who.int/news-room/fact-sheets/detail/human-rights-and-health>.

Zubrzycki, J., Green, S., Jones, V., Stratton, K., Young, S. and Bessarab, D. (2014). Getting it Right: Creating Partnerships for Change. Integrating

Aboriginal and Torres Strait Islander Knowledges in Social Work Education and Practice, Australian Catholic University, Research Bank. Available from: <http://www.acu.edu.au/__data/assets/pdf_file/0010/655804/Getting_It_Right_June_2014.pdf>.

https://doi.org/10.1142/9781786348579_0002

CHAPTER TWO

Narrative Ethics in Indigenous Health

Alireza Bagheri,* Crystal Pirie† & Sareh Bagherichimeh‡

Indigenous health narratives...

— *A nurse checked my arms to see if I have been using intravenous drugs.*
— *I am not even comfortable to introduce myself as an Indigenous nurse in my hospital.*
— *I feel my rights as a human being are being ignored.*
— *Racism forces me to constantly explain that I'm not a burden on society.*
— *I do not get enough medical attention because I am not like Them.*

Abstract

Narrative ethics emphasizes both the importance of giving patients an opportunity to tell their life stories and the importance of listening to those stories. Narratives can illustrate the disadvantages felt by minority groups, who are suffering from racism and discrimination in healthcare systems. They also can help to identify barriers in providing care

*Research Affiliate, Center for Healthcare Ethics, Lakehead University, Canada.

†Senior Director of Indigenous Collaboration, Thunder Bay Regional Health Sciences Centre, Canada.

‡Research Assistant, Institute of Medicine, University of Toronto, Canada.

without discrimination. Indigenous narratives are a source of insight into a health experience influenced by colonialism and socio-economic contexts. They shed light on facts other than those described in statistics on mortality rates and disease incidence among Indigenous peoples.

By emphasizing the importance of narratives in healthcare, this chapter provides several Indigenous narratives which attracted nationwide attention in Canada. These narratives illustrate the continuation of racism and discrimination in Indigenous communities and our failure to eliminate injustice in Indigenous health.

Introduction: The Importance of Narratives in Healthcare

Caring for patients involves an understanding of and a response to their needs. An ethics of care focuses on the moral significance of interpersonal relationship that develop between people, particularly when one person is vulnerable because of sickness or age, and it values as morally significant those virtues associated with care [Nicholas & Gillett 1997].

As Jan Paulsen [2011] argues, caring involves working with narratives, and it demonstrates how this can be done. A narrative approach is therefore intrinsic to an ethics of care. It generates insights and an ability to discern phenomena that other forms of ethics either lack or would be hard pressed to incorporate. Joan Tronto [2004] also suggests that the narratives require conversation, listening, interpretation, and what other commentators have referred to as attentiveness to and knowledge of the other's needs. In addition, narrative ethics recognizes the importance of temporality. As Porter Abbot states in his work *An Introduction to Narrative* [2014] narrative is "the representation of an event or a series of events".

Narrative medicine is medicine practiced with the competence to recognize, absorb, interpret, and be moved by the stories of illness. In medicine, narratives help healthcare professionals to recognize patients and diseases, convey knowledge, accompany patients through the ordeals of sickness and can ultimately lead to more humane, ethical, and effective healthcare [Charon 2006]. In healthcare ethics, a narrative has two phases, the telling of the story, and the hearing of the story. However, in many situations the stories that are being told are not necessarily heard.

Rather than trying to find an approach which escapes from or removes the specifics of culture, history or relationship, a narrative ethic seeks to make these particulars visible. Thus, narrative ethics recognizes the centrality of a patient's story, and grapples with the specifics of context and relationship. In addition, narrative ethics recognizes that questions of interpretation and power are a necessary part of ethics [Nicholas & Gillett 1997].

Narrative is integral to many parts of medicine. Story forms the basis of medical care in the narratives that patients bring to their doctors and in the narrative the doctor constructs in relation to the patient. Narratives also structure the conversations between health professionals and provide a major vehicle for the transmission of knowledge and the formation of a professional relationship. Clinical choices are not isolated from all else that happens in people's lives but are part of an ongoing narrative. Narratives can illustrate disadvantages of minority groups, who are suffering racism and discrimination in healthcare system.

Narratives of Indigenous Health: Less Heard

A narrative ethic recognizes that the voices of the most vulnerable and those on the margins of a society, in their narratives and interpretations, have something to offer [Nicholas & Gillett 1997]. In the last four decades, the stories of those people who suffered under colonial policies have shed lights on the relation between colonization and well-being of Indigenous people around the world. As observed by Reading & Wien [2009], there is increasing consensus that the health inequalities, disproportionate rates of disease, disability, addiction and violence in Indigenous communities around the world are the result of colonization. Therefore, information about Indigenous health cannot be understood outside of the context of colonial policies and practices both past and present [Allan & Smylie 2015].

Colonization has been recognized internationally as a key determinant of health for Indigenous people, including by the World Health Organization. Another study showed the link between colonization, racism, social exclusion and determinants of health including healthcare, education, housing, employment, income, food security [Loppie *et al.* 2009].

By providing narratives of the lives of three elderly Sami women, a study in Norway illuminates how the history of colonization is present in elderly women's lives and impacts their health experiences. The authors argue that understanding health as a condition of subjectivity and as influenced by broader historical and social contexts are essential to gaining a richer understanding of the health of Indigenous people [Blix *et al.* 2012]. A narrative ethic asks important questions: whose story is being told, and by whom? Whose interpretive framework is being given authority? How do those of us with social or institutional power respond to narratives with which we are uncomfortable, or which challenge our position [Frank 1994]? The narratives of life in a residential school in Canada and Norway for instance, are not just the stories about the past. The experience of discrimination and stigmatization still live in Indigenous peoples' lives. As Blix *et al.* [2012] argue, the narratives of Indigenous people illustrate that they are not necessarily passive victims of the legacy of colonialism; on the contrary, they are expressions of the agency of "the oppressed".

In response to patients' narratives in Indigenous communities, several international instruments have been developed to address racism, discrimination and inequities in providing healthcare to Indigenous communities. For instance, in 2007, the United Nations Declaration on the Rights of Indigenous Peoples was adopted by the General Assembly. The Declaration [UN 2007] is the most comprehensive international instrument on the rights of Indigenous peoples. As discussed in the previous chapter, it establishes a universal framework of minimum standards for the survival, dignity and well-being of the Indigenous peoples of the world and it elaborates on existing human rights standards and fundamental freedoms as they apply to the specific situation of Indigenous people. However, the question remains as to why these international instruments have not yet been successful in removing racism, discrimination and health inequities in Indigenous communities. There is a hope that patient narratives can illustrate the continuation of racism and discrimination in Indigenous communities despite the existing international instruments and our failure to eliminate injustice in Indigenous healthcare. Narratives can also help us to identify barriers in providing care based on these international recommendations as well as human rights.

Canadian Indigenous Health Narratives

In what follows we discuss narratives about Indigenous patients which were published in the media and attracted national attention in Canada. These narratives show how systematic racism, discrimination, and sometimes lack of empathy and compassion among healthcare providers resulted in the deaths of Indigenous patients.

First narrative: Racism in healthcare systems can be fatal

"The effects of systemic racism are pervasive in Indigenous communities and racism in the Canadian healthcare system can be fatal." This is a statement by the Indigenous Health Working Group of the College of Family Physicians of Canada and Indigenous Physicians Association of Canada. The Working Group reported on a narrative about a patient who died while waiting at the emergency room for 34 hours without ever receiving treatment.

Maclean's, a national magazine in Canada reported that a 45-year-old double-amputee Indigenous patient spoke to a triage aide upon his arrival at the emergency room in a hospital and then wheeled himself into the waiting room where he languished for the next 34 hours [see Puxley 2013]. The patient was referred to the hospital by a community physician for a bladder infection. While at the emergency room, the patient vomited on himself several times, and other visitors pleaded with nurses and security guards to attend to the patient. Following the 34-hour wait, the patient died of the bladder infection in the waiting room without ever receiving treatment. The patient's family, their legal counsel, and local Indigenous leaders asked a provincial inquest into the matter to strongly consider the ways in which the patient's race, disability, and class resulted in his lack of treatment and subsequent death. However, in February 2014, the Sinclair family withdrew from the provincial inquest due to frustration with its failure to examine and address the role of systemic racism, and in the treatment of Indigenous patients. By emphasizing that the systemic racism leads to health inequities, the working group of the College of Family Physicians of Canada and Indigenous Physicians Association of Canada has provided a set of recommendations for family physicians in how they can build trust and form lasting relationships with Indigenous patients to prevent the tragedies happening in caring for Indigenous patients.

Second narrative: Turning away intoxicated patients

In 2012, Ina Matawapit of North Caribou Lake First Nation died in police custody [Wilson 2018]. *The National Post* in Canada [Martell & Nickel 2019] reported that after being arrested for public intoxication, the police drove Marawapit to a federal government-run clinic as she had suffered from a blow to the head and was complaining of chest pains. A nurse assessed the semi-conscious Matawapit in the back of the police car and according to testimony at the 2018 inquest, the nurse told the officer to bring her back when sober. However, on the drive back Matawapit's condition deteriorated quickly. Unable to find a pulse, the police sped back to the clinic where they were unable to resuscitate her, partly due to a faulty defibrillator and shortage of oxygen supply, Matawapit died of heart failure [Martell & Nickel 2019].

There was an inquest into Matawapit's death, and it is important to note that it was due to the Coroners Act [1990] that mandates inquiry into death while in police custody. Otherwise it is likely that her death would have gone unexamined given that the federal government has not investigated other cases at clinics in the Indigenous reserves for at least the past nine years [Wilson 2018]. This suggests that there may well be many more cases that have gone unexamined. The inquest juries made a total of 27 recommendations to six different agencies and facilities aimed at preventing similar deaths. However, the majority of recommendations were made to Indigenous Services Canada. Thirteen recommendations were directed to healthcare providers, seven involved equipment, four involved education or training, five involved checks and balances, and two were specifically related to intoxicated individuals. The broadness of the range and the scope of recommendations as well as the number of different agencies and facilities the recommendations were directed to, speak of how the issues surrounding poor outcome cases in the Indigenous communities are systemic [Wilson 2018].

This tragic case also brought attention to the "Northern Protocol", a term used in the coroner's verdict and by the attending nurse in her testimonial at the inquest, referring to the practice of turning away intoxicated patients in the northern reserves [Prokopchuk 2018]. After the case of Tracy Okemow in God's Lake Narrows Manitoba five month later, who was also sent back from a clinic to sober up in a cell, the federal government enacted a policy deeming detention of intoxicated patients in cells as inappropriate [Martell & Nickel 2019]. The inquest jury also made recommendations in regard to care of intoxicated individuals and called for the development of protocols for assessment and management of intoxicated patients [Wilson 2018].

Third narrative: Suicide: a repeated narrative[1]

Greg died as he lived, by his own rules. At the age of 32 he made the decision to end his life, believing that his family and friends would be better off without him. Greg was born in Burnaby, British Columbia, March 18, 1987. His mother, Kim, believes he was her birthday present as her birthday is March 22. He had a good, healthy childhood and grew up to be blessed with a lot of love and happiness in his life, including his wife Lichelle, and his two daughters, Nova and Cedar. In his career, he enjoyed working with people, helping them to improve their health status and outcome and provided supportive services to Personal Support Workers.

Greg was smart and logical and was also gifted with foresight. A gift that, at times, scared his family but is highly valued in the Indigenous community, especially as "Mushkiki Mukwa" or Medicine Bear, Greg's traditional name. His gifts were so evident that his parents were approached on two occasions by two different traditional medicine men asking permission to mentor Greg in traditional medicine. While in grade four at school, he was diagnosed with Phonetic Logical Learning Disability. He was also tested for metal toxicity and his results showed that he had a high level of copper in his body. A year later, at the age of 10, he was diagnosed with Attention Deficit Hyperactivity Disorder (ADHD). At the time, his parents made the decision not to medicate Greg with Ritalin, believing the medication to be too new. The decision to begin the use of Ritalin, six years later, was made because Greg started cutting

---

[1] Greg's wife and parents have provided signed consent to share Greg's story.

himself. They thought a medication regimen would work because Greg was always compliant to requests, and was eager to participate in activities, especially those involving his dad. Unfortunately, two months later Greg began to snort his medication and experienced hallucinations that were so severe his parents had to have a friend sit with Greg throughout the day while they were at work. The family pursued counselling services for Greg around this time.

At the age of 14, Greg began using marijuana and continued to do so until his untimely passing. He also introduced marijuana to his brother who then had an unexpected adverse reaction while smoking with friends. It was so bad he was brought to the hospital, but the symptoms he experienced did not end when the effects of the marijuana wore off. He was eventually diagnosed with schizophrenia, and Greg blamed himself for his brother's diagnosis. Greg suffered from compound headaches and chronic facial nerve pain from a workplace accident, which resulted in many visits to the doctor. He was accused of being "drug seeking" despite it being noted in his file that he suffered from chronic pain. Greg was a firm believer that everyone was responsible for their own health and well-being, including himself. In the weeks leading up to his death, he tried very hard to control the situation and sought help. He reached out to and had the support of friends and family, visited a walk-in clinic, was admitted to the hospital, and attended a scheduled appointment with his family physician. However, it was during this period of his life that Greg began to doubt his self-worth and believe that he was a burden to society.

In May 2019, Greg's friend called the police to help get Greg to the hospital since he knew Greg was in a bad place and wanted to end his life. He was met by his wife at the hospital, where he was triaged, assessed and in the process of being admitted for being a danger to himself. Greg was to be transferred from the Emergency Department to the Adult Mental Health unit as soon as a bed became available. Prior to a bed becoming available for Greg there was a shift change. The incoming physician met with Greg and made the determination that he was safe to return home. Greg's wife advised the new physician that this was not a good or safe plan and made it clear that she did not agree. Despite this, Greg was released with a discharge plan to follow up with his family doctor and access community-based resources. Shortly after the discharge, Greg and his wife determined he could not wait for his appointment to see his family doctor, so they visited a walk-in clinic. The physician at the clinic advised Greg that he must see his family doctor. Eventually, Greg finally managed to see his family doctor of approximately 20 years who then advised him to go back to the Emergency Department as he needed immediate intervention. Greg shared his recent, negative experience at the hospital with his doctor and then ran from his office, leaving his wife behind. His doctor ran after him and is one of the last known people to see him alive.

Upon the doctor's return to his office he called the police for assistance, fearing Greg would harm himself. This incident marked the last time Greg would be rejected or denied help. His body was found the next day. According to Lichelle, Greg felt like he wasn't being heard. She believes that the system failed her husband. Miserably. His parents are left to wonder if things could have been different for their son if the medical system was structured to look at the whole person and not address one ailment in and of itself. Although Mushkiki Mukwa has gone to the spirit world the family continues to wonder what more they could have done, where else they could have taken him for the help he was looking for, for the help he needed.

Fourth narrative: Apathy and lack of care

In 2010, Romeo Wesley died while in custody at Cat Lake First Nation Nursing Station. Jody Porter at CBC News [2017] reported that Wesley arrived at the nursing station in a panic, having taken oxycodone and cannabis. His outbursts resulted in the police being called. By the time the police arrived, Wesley was confined in the hallway after punching a glass window, assaulting the security guard and sustaining cuts on his arm. The police cuffed his hands and had him lie face down with each having a foot on his back applying pressure while Wesley's legs kicked up and down. It took eight minutes after he stopped kicking for a doctor to check his vitals and perform CPR but he never recovered. According to the autopsy report Wesley died from a combination of "chest compression with positional restraint", agitation and trauma in a man with KCNJ2 gene mutation (is characterized by periodic paralysis and cardiac arrhythmias) and severe alcohol withdrawal [Cameron 2017]. The mandatory inquest under the Coroner's Act was originally scheduled to be held in Sioux Lookout, Ontario, but Indigenous leaders fought to have the proceedings take place in the fly-in community of Cat Lake where Wesley resided, and the incident took place. In the affidavit to motion for the change of venue the Cat Lake Chief Russell Wesley argued "The traditions of Aboriginal people are the same as Euro-Canadian legal principles in recognizing that investigation and inquiry, judgement or punishment, out of sight of those involved, does not serve the communities' interest" [Porter 2015].

At the inquest seven years after Wesley's death, the jury ruled his manner of death accidental and made 53 recommendations to nine different agencies to prevent similar deaths in the future. [Cameron 2017]. The issue of lack of communication and protocols became apparent in the inquest. The police officers testified that after restraining Wesley they were waiting on instructions from the medical staff at the nursing station and that they were unaware of his medical condition. The officers believed the medical staff to be in charge of their patient's well-being as they were in "their house", meaning the nurses station. While Dr. Harriet Lennox testified that "I assumed that they had police protocols and would know when it was safe to do those things." The 45-minute security video from the nursing station was released despite the bid by the Ontario coroner's counsel to keep it from public view. It demonstrates the apathy and lack of care Mr. Wesley received the night of his death. The security video captures a doctor watching as police officers stepped on Mr. Wesley's back and a nurse was visible at various times mopping the floor and offering towels to police officers.

Conclusion

Narrative ethics has two roles, telling the story and listening to the story. In Indigenous health, narratives have been told repeatedly, now it is time for us — the public, healthcare providers, health policy makers and politicians — to listen to these narratives and act, and to change the picture in a morally acceptable manner which recognizes the ongoing effects of racism and colonialism.

The international instruments have not yet been successful in removing racism, discrimination and health inequities in Indigenous communities. It is, therefore, important to take action that can change the lives of Indigenous people around the world. This is one important role for patient narratives: to illustrate these issues and to help us to identify barriers in providing care based on these international recommendations as well as human rights.

References

Abbot, P. (2014). *The Cambridge Introduction to Narrative* (Cambridge University Press, Cambridge).

Allan, B. and Smylie, J. (2015). First Peoples, Second Class Treatment: The Role of Racism in the Health and Well-Being of Indigenous Peoples in Canada (Wellesley Institute, Toronto).

Blix, B. H., Hamran, T. and Normann, H. K. (2012). Indigenous Life Stories as Narratives of Health and Resistance: A Dialogical Narrative Analysis, *Canadian Journal of Nursing Research* 44(2): pp. 64–85.

Cameron, D. (2017). *The Coroners Act — Province of Ontario*, Verdict of Coroner's Jury Office of the Chief Coroner, Cat Lake First Nation, ON. Available from: <https://www.mcscs.jus.gov.on.ca/english/Deathinvestigations/Inquests/Verdictsandrecommendat ions/OCCInquestWesley2017>.

Charon, R. (2006). *Narrative Medicine* (Oxford University Press, New York) p. 266.

Coroners Act, R.S.O. 1990, c. C.37.

Frank, A. (1994). Editor's Introduction: Case Histories and the Ethics of Voice. *Second Opinion* 17(3): pp. 44–48.

Martell, A. and Nickel, R. (2019). Deaths, Bad Outcomes Elude Scrutiny at Canada's Indigenous Clinics, *National Post*, October 24. Available from: <https://nationalpost.com/pmn/health-pmn/deaths-bad-outcomes-elude-scrutiny-at-canadas-indigenous-clinics>.

Nicholas, B. and Gillett, G. (1997). Doctors' Stories, Patients' Stories: A Narrative Approach to Teaching Medical Ethics, *Journal of Medical Ethics* 23, pp. 295–299.

Paulsen, J. E. (2011). A Narrative Ethics of Care, *Healthcare Analysis* 19: pp. 28–40, DOI: 10.1007/s10728-010-0162-8.

Porter, J. (2015). 'Horrified' First Nation Chief asks for Change in Venue for Inquest, *CBC News*, October 14. Available from: <https://www.cbc.ca/news/canada/thunder-bay/horrified-first-nation-chief-asks-for-change-in-venue-for-inquest-1.3268974>.

Porter, J. (2017). Death of First Nations Man Handcuffed and Stepped on by Police was Accidental, Inquest says, *CBC News*, July 26. Available from: <https://www.cbc.ca/news/canada/thunder-bay/romeo-wesley-verdict-1.4221069>.

Prokopchuk, M. (2018). Inquest into North Caribou Lake Woman's Death Brings Answers, Hope for Change, Councillor says, *CBC News*, February 21. Available from: <https://www.cbc.ca/news/canada/thunder-bay/ina-matawapit-inquest-grace-1.4543535>.

Puxley, C. (2013). Woman in ER Where Man Died After Lengthy Wait Says It Was Obvious He Needed Help, *Maclean's Magazine*. Available from: <https://www.macleans.ca/general/woman-in-er-where-man-died-after-lengthy-wait-says-it-was-obvious-he-needed-help/>.

Reading, C. L. and Wien, F. (2009). *Health Inequalities and Social Determinants of Aboriginal Peoples' Health,* National Collaborating Centre for Aboriginal Health (Prince George, BC: National Collaborating Centre for Aboriginal Health).

Tronto, J. (2004). What Can Feminists Learn About Morality from Caring? In Sterba, J. B. (ed.) *Ethics: The Big Questions* (Blackwell Publishing, Oxford).

United Nations (2007). UN Declaration on the Rights of Indigenous Peoples. Available from: <https://www.un.org/development/desa/indigenouspeoples/wp-content/uploads/sites/19/2018/11/UNDRIP_E_web.pdf>.

Wilson, M. (2018). *The Coroners Act — Days Inn*, Verdict of Coroner's Jury Office of the Chief Coroner, 3 Sturgeon River Road, Sioux Lookout.

CHAPTER THREE

Indigenous Bioethics in Brazil: Promoting and Protecting Indigenous Peoples

Leo Pessini, * *Anor Sganzerla*† *& Diego Carlos Zanella*‡

Abstract

The social perspective adopted by Latin-American bioethics and the distinctive features developed in Brazil, particularly a bioethics of protection, contribute to a deepening of concern about Indigenous peoples. Based on analysis of legislation in context this chapter aims to analyze how bioethics can contribute to the promotion and protection of Indigenous peoples in Brazil. The concepts that underlie the arguments refer to the *Universal Declaration on Bioethics and Human Rights* [UNESCO 2005] and the *Declaration on the Rights of Indigenous Peoples* [UN 2007], as well as other authors who work with social and pluralist perspectives.

*We have co-dedicated this chapter to Leo Pessini, who passed away in 2019. He was Camillian Religious Minister of the Infirm and General Moderator of the Camillianum: International Institute of Pastoral Health Theology.
†Professor of Bioethics Graduate Program, Pontifical Catholic University of Paraná (PUCPR), Curitiba, PR, Brazil.
‡Professor of Humanities and Languages Teaching Graduate Program, Franciscan University (UFN), Santa Maria, RS, Brazil.

Introduction

Bioethics can be defined as the science of human survival, as a guarantor instrument of the future, linking ethics with the need for preservation of life [Potter 1971]. It was born to ensure that scientific and technological research maintains respect for basic principles, such as respect for human rights, human dignity and fundamental freedoms [UNESCO 2005], as well as the principles of autonomy, beneficence, non-maleficence and justice, described in the *Nuremberg Code* of 1947, the *Helsinki Declaration* of 1964 and its revisions, and the *Belmont Report* of 1978. These are discussed by Beauchamp and Childress in the book *Principles of Biomedical Ethics* [2012].

The *Universal Declaration of Human Rights* [UN 1948] opened up the internationalization of human rights, promoting the elaboration of many international protection documents and the observance of these by states, to the detriment of national sovereignty in the search for the welfare of all peoples regardless of borders. The 2005 the *Universal Declaration on Bioethics and Human Rights* was formulated to ensure that individuals or groups of individuals had their essential rights protected in the face of scientific and technological research, highlighting Indigenous peoples as potential targets for such research.

Considered to be vulnerable groups in the face of scientific practices, Indigenous peoples are the subjects of protection that recognize human dignity and embrace international human rights documents.

Indigenous communities have a wide range of knowledge, which has been maintained and developed over time. It has served to maintain their cultures and the balance of nature since it comprises important practices related to the environment and biological diversity. Although Indigenous communities are of interest to researchers, the question remains, who can develop scientific practices in these communities? If legal and ethical research is not ensured, research can prove destructive to the culture and maintenance of these peoples.

The model of bioethics based on clinical ethics is insufficient for the reality of the Latin American continent [Pessini 2016; Pessini *et al.* 2010]. Therefore, scholars began to seek a bioethics that had deeper effects in the underdeveloped reality of Latin America, marked by social inequality, exclusion and injustice. A bioethics of protection recognizes inequalities that damage the social structure and is concerned with populations that suffer from restrictions of freedom due to politics, lack of empowerment, and a predisposition to increased susceptibilities to harm [Schramm 2008; Schramm & Kottow 2001]. Therefore, it is imperative that bioethics and human rights are observed in any research that may be carried out with individuals and/or groups considered vulnerable. This approach contributes to the preservation of traditional knowledge and the return of the benefits generated to human dignity and freedom. Thus, the purpose of this chapter is to analyze how bioethics can contribute to the promotion and protection of Indigenous peoples in Brazil.

Indigenous Peoples in Brazil

The vast majority of Brazilians still ignore the immense diversity of Indigenous peoples and cultures living in their country. At the time of discovery by the Portuguese in the 15th century, it was estimated that there were more than a thousand Indigenous tribes and cultures in Brazil, ranging between 2 and 4 million people. Today, in the 21st century, there are approximately 250 Indigenous tribes who speak more than 150 different languages. Indigenous people now number approximately 900,000, live in rural areas or in cities and comprise approximately 0.47% of the total population of the country. Most of this population is distributed in several villages, located in the interior of more than 700 Indigenous territories.[1]

Today, speaking about Indigenous peoples in Brazil implies recognition that: (i) there were already human populations occupying specific territories of what is now known as Brazil, which was discovered on April 22, 1500, by Pedro Álvares Cabral (1467/8–1520); (ii) the origin of these

---

[1] All data presented in this section correspond to the last Population Census, conducted by the Brazilian Institute of Geography and Statistics (IBGE, acronym in Portuguese), in 2010. Information was collected from the following websites: (i) IBGE: Indigenous Peoples: <https://indigenas.ibge.gov.br>; (ii) Indigenous Peoples in Brazil: <https://pib.socioambiental.org/pt/Página_principal>; (iii) Social and Environmental Institute (ISA, acronym in Portuguese): <https://acervo.socioambiental.org>. Date of access: April 16, 2018.

human populations, called natives or Indigenous, is controversial because they were already inhabitants of that territory before the European occupation; (iii) some groups of people who currently live in Brazil are historically linked to these first people; (iv) the Indigenous peoples living in Brazil today have a long history that, before discovery, is different from, but after discovery, is similar to the history of Western civilization; (v) like all human groups, Indigenous peoples have cultures that result from the history of the relationships among human beings themselves and between them and the environment, a history that has been (and continues to be) drastically altered by the reality of colonization; (vi) territorial division in countries does not necessarily coincide with the Indigenous occupation of land. In many cases, the people who now live in a region of international borders already occupied that area before the creation of divisions between countries. That is why it makes much more sense to speak of Indigenous peoples in Brazil than of Brazil itself [Tiradentes & Silva 2008].

Brazil has immense ethnic and linguistic diversity, which coexists with national society in all regions of the country, in a true human laboratory of contact, cultural exchanges and challenges. In this sense, it is necessary to recognize and value the differences about the planet on which one lives and where life exists through the phenomenon of diversity. At the cultural level, it is important to say that life is much deeper and richer, as a result of coexistence with different people and cultures.

The Denial of Indigenous Peoples' Identity

Coverage of Indigenous people in the Brazilian media generally gains attention when there is a conflict such as a territorial dispute between farmers and Indigenous peoples. These clashes are often violent and result in fatalities. Those responsible for the massacres often go unpunished because the public authorities have no interest in investigating the deaths of "worthless" people as well as the poor or other social groups classified as vulnerable.

In recent years, however, some reports, documentaries and stories about Indigenous peoples have attracted attention, as they seek to propagate the idea that they are moving away from their more traditional way of life. These accounts strongly emphasize the use of information technologies such as mobile phones by Indigenous peoples.[2] Behind this propaganda, which shows a de-characterization of Indigenous culture, are the interests of public authorities and especially of agribusiness. Insofar as native peoples "are no longer considered Indigenous," it is no longer necessary to assure them the rights guaranteed by the Federal Constitution [Brazilian Federal Constitution 1988], which guarantees perpetual usufruct by Indigenous peoples of their lands and reserves.

Thus by taking possession of the present Indigenous reservations, they are denying them their identity without breaching the Federal Constitution, which requires the denial of the identity of the Indigenous peoples as Brazilian citizens.

The justifications used to deny the identity of Indigenous peoples are based on diverse arguments, especially the change of dress, access and use of information technologies, use of State resources for health issues, such as vaccination, access to water and power, the use of the Portuguese language for communication.

This way of thinking, conveyed by the media, does not allow that Indigenous peoples can integrate with other societies, and as a consequence, exchange experiences, cultures and even resources. It has been said that for Indigenous people to maintain authenticity, they must keep themselves away from other civilizations, thus denying historical practice, which has always involved cross-cultural influence, whether in the field of art, science, religion, technology or customs. This way of thinking is detrimental both in denying the resources and possibilities of modern societies to Indigenous peoples and in denying the wisdom of Indigenous cultures that could contribute much to modern societies [Tiradentes & Silva 2008].

---

[2] The Brazilian media have portrayed Indigenous populations as if they were losing their origins because of the use of new digital technologies, as can be seen in the report in Portuguese on the effects of the use of technology on Indigenous life. Available on:<http://g1.globo.com/ap/amapa/noticia/2014/05/avanco-da-tecnologia-em-aldeia-muda-cotidiano-de-indios-no-amapa.html>. Accessed April 30, 2017.

This type of mentality is also reinforced in children's books used in schools whenever they show images of Indigenous people during the discovery period, living in the middle of the forest with a bow and arrow with many naked children in huts made of straw, bathing in a river and without any kind of modern resources. In this view if Indigenous peoples wish to maintain their identity, they should continue their traditional ways of searching for food, housing, health protection, and religious rituals. This kind of teaching, which begins in childhood for most students, ends up becoming the most habitual way of thinking about Indigenous people in adult life.

Indigenous cultures, like all cultures, change according to interests and needs. Cultures are not "museum pieces" that are unchangeable or untouchable. They are a living part of society that represent ways of being, acting and thinking. And it is precisely this capacity for adaptation that will ensure future survival. To respect the specifics and transformations of each culture requires the ability to understand the concept of difference, since no one can accept what they do not understand. And in this sense, Western culture has shown its difficulties in accepting different cultures, which ends up impoverishing both native and modern cultures. Hence, promoting dialogue between different cultures and moralities is a challenge to today's bioethics.

Respect for Millennial Knowledge in Health

The cultural diversity of the Brazilian Indigenous peoples is very significant, which means they have different ways of understanding, for example, life and death, health and disease as well as transcendence, nature and birth. At the root of these different understandings is a religious structure marked by strong symbols, both to deal with everyday issues and to address health and life in general.

Among Indigenous groups, the people who take responsibility for the health of their community base their knowledge about health in a tradition passed down from generation to generation, which assures a specificity in relation to how they perceive health and disease. In spite of this, Kleinman states that the way health and disease is perceived by Indigenous peoples has very similar elements to those seen in Western culture, such as: (i) cultural understanding of the disease; (ii) the construction of strategies in the view of healing practices; (iii) the demonstration of practices that prevent, ameliorate or worsen disease; and (iv) the administration of therapeutic results [Kleinman 1978]. With this in mind, we shall not dwell on what is similar in different cultures, but rather on the specifics of the Indigenous traditions.

Among the specific issues that are likely to be recognized is how the idea of disease is understood among Indigenous peoples, since it is believed that both gods and spirits are responsible for the illness as well as for the cure of the individual. Therefore, the search for the cure of diseases is linked to the transmission of empirical knowledge, which is passed on from generation to generation, or even through mystical inspirations of the traditional knowledge of shamans, who have access to the spiritual world and from where they can extract a position or a therapy for the sick, whether in the form of religious rites, with songs or dances, or in the form of medicines extracted from plants.

In regard to knowledge, and as a mark of this process, Dantas [2006] highlights the collective and shared structure, which is the result of practices and experiences related to social and cultural spaces as well as customs and traditions.

At present, however, Indigenous peoples also have access to modern healthcare. But this possibility, in turn, does not rule out the traditional way of thinking about the health/disease relationship of the community. One of the challenges is to admit the effectiveness of scientific knowledge on the modern health system, and not to neglect traditional knowledge.

The World Health Organization (WHO), in the document *WHO Traditional Medicine Strategy: 2014–2023* [2013], addresses traditional health systems and their preventive practices, their diagnoses, as well as their spiritual and bodily therapies with the use of plants and animals. It also highlights that the empirical and symbolic efficiency of such practices has to be properly incorporated into modern health systems, as a joint articulation of knowledge, emphasizing the well-being and the safety of the sick and all those around him/her.

An example presented in a national magazine shows how it is possible to bring together traditional knowledge with modern practices in order to recover the health of the patient.[3] In the year 2009, an Indigenous child, who had been attacked by a snake, was admitted to a hospital in the city of Manaus. The child underwent several surgeries to extract the necrotic tissue. However, a serious lesion to the right foot required it to be amputated. On learning of the recommendation to amputate, the father of the child asked the shaman if he could apply the traditional remedies to the affected area and perform prayers and rites according to the traditional custom. Since the father's request was denied by the hospital's management, he appealed to the Attorney General's Office, which authorized him to remove the child from the hospital and to take him to an Indigenous place of healing, where he would receive the care of his shaman. After a few days, the director of the hospital sought out the father of the child to propose a joint treatment, that is, modern medicine and traditional knowledge. The father accepted the proposal and the child was accommodated in an ICU, where, in addition to receiving snake antivenom serum and antibiotics, he also received a visit from the shaman who performed religious rites and healing practices with the use of herbs.

This case is interesting because the hospital accommodated the requests made by the shaman himself, in particular, a request that the nurses attending to the child were not pregnant or menstruating nor had sexual intercourse in the last 24 hours. After a few days, the condition of the child was much better, and amputation of the foot was no longer necessary.

This articulation between traditional and modern knowledge, as presented in the case above, unfortunately does not constitute a respected practice in a daily hospital routine, which makes bioethics an even more indispensable tool, guaranteeing dialogue, respect for cultural and religious differences, and the acceptance of the other in such differences.

Ethics in Research

The vulnerability of Indigenous peoples in the face of scientific research and practice has required the need for international protection, as can be

---

[3] Available on: <http://revistaepoca.globo.com/Revista/Epoca/0,,EMI62314-15228,00-JUNTOS+MEDICOS+E+PAJE+EVITAM+AMPUTACAO.html>. Accessed April 30, 2017.

seen in the *Universal Declaration on Bioethics and Human Rights*, which states in its article 8 — *Respect for human vulnerability and personal integrity* — that "(i)n applying and advancing scientific knowledge, medical practice and associated technologies, human vulnerability should be taken into account. Individuals and groups of special vulnerability should be protected, and the personal integrity of such individuals respected".

Article 6 in the above document, which deals with consent in research, states that: "Any preventive, diagnostic and therapeutic medical intervention is only to be carried out with the prior, free and informed consent of the person concerned, based on adequate information". It is complemented by article 11, that emphasizes that all forms of discrimination and stigmatization are condemned because they violate human dignity, human rights and fundamental freedoms. Regarding the results of scientific research, article 15 predicts that benefits, derived from any scientific research and its technological applications, should be shared with the individuals or groups involved.

In Brazil, research on human beings is regulated by resolutions 466/2012 and 510/2016 of the National Health Council.[4] The first presents the norms that regulate research involving human beings while the second presents norms applicable to research in the areas of human and social sciences, whose methodological procedures involve the use of data directly obtained from the participants or information that is identifiable. However, the concern about the vulnerability of Indigenous peoples in relation to scientific research dates back to the 1990s, when Brazil created its first ethical regulation for research involving human beings, that is Resolution 196/1996 of the National Health Council. Thus, from that moment on, discussions about ethics in research with human beings, and especially with Indigenous populations, started to have new points of reference. The progress of this debate continued with Resolution 304/2000 of the National Health Council, which presents the specific norms for research involving Indigenous populations. Through these resolutions, any research on human beings began to be regulated, and particularly research involving Indigenous populations.

---

[4] All resolutions of the National Health Council can be accessed through the link: <http://conselho.saude.gov.br/resolucoes/reso_inicial.htm>. Accessed April 30, 2018.

These regulations define the methods and techniques of research and produce knowledge about Indigenous worlds, defining it as vulnerable.[5] One of the issues is that the resolution is based on a biomedical perspective, which structures health policies in Brazil. This standpoint universalizes the ethical problems from the biomedical area to other areas of knowledge, thus revealing the hegemony of biomedical knowledge [Pessini 2016; Pessini et al. 2010]. On the other hand, it is worth noting that this legislation was a great advance in the organization of the protection system of persons exposed to biomedical research.

In this context, attention should be drawn to the situations of vulnerability that characterize the *modus vivendi* of many groups of people in Latin America, among whom are Indigenous peoples. However, in the face of vulnerability, which is fundamentally characterized as lack of protection, exploitation, or denial of identity for Indigenous peoples, an individualistic bioethics does not help protect the vulnerable as a group or community [Pessini 2016; Pessini et al. 2010]. Social, political and economic conditions accentuate these inequalities still further. In this regard, the reality of Latin American bioethics demands an ethical and social perspective, with a strong concern for the common good, justice and equity rather than a perspective of individual rights and personal virtues [Schramm 2008; Schramm & Kottow 2001].

Bioethics of protection recognizes inequalities that damage the social structure and is concerned with populations, including those that suffer from restrictions of freedom due to deprivation, lack of empowerment, and predisposition to increase susceptibilities. Schramm and Kottow [2001, 2008] suggest a "Bioethics of Protection" as a theory that reflects and provides guidance on (public) biomedical practices as they occur in marginalized societies, with clear considerations about the visions of ethics in biomedical protection practices to meet specific demands and recognition of outdated ethics in political, social and philosophical life. A

---

[5]The National Health Council (CNS, acronym in Portuguese) is the highest decision-making body of the Unified Health System (SUS, acronym in Portuguese) and is linked to the Ministry of Health. It is the responsibility of the council to approve the health budget and monitor budget execution. The council is also responsible for approving the National Health Plan every four years.

bioethics of protection considers that the government is committed to protecting all members of society in the face of any intervention that is not stable regardless of State structure as well as the awareness of vulnerability as a human condition. It also takes into account that social philosophy and policy should ensure the protection of citizens against violence, poverty and any kind of violation of human rights [Schramm 2008; Schramm & Kottow 2001].

Government Negligence

The Brazilian Congress is currently dominated by different groups, including lobbyists from the agricultural sector and large landowners, which results in the protection of Indigenous land being under constant threat.

The Brazilian Federal Constitution is very clear when it refers to the defense of lands and Indigenous reserves. Article 231 of the Constitution recognizes the Indigenous peoples' "original rights over lands they traditionally occupy" [Brazilian Federal Constitution 1988]. Paragraph 4 states that "the lands referred to in this article are inalienable and indisposable and the rights thereto are not subject to limitation". Paragraph 6, in turn, provides that "acts with a view to occupation, domain and possession of the lands referred to in this article are null and void, producing no legal effects."

Although the law is very clear, in practice the invasions of Indigenous lands in order to expel the owners and to take over the existing wealth, be it the wealth of the wood of the forests or mining, is still a practice that has not been curbed by the public authorities. The demarcation of land itself to Indigenous communities provided for in the 1988 Constitution has not been carried out as yet. The Constitution says: "The union will conclude the demarcation of Indigenous lands within five years from the promulgation of the Constitution".

This omission by the State is intentional, and occurs because there is pressure from those who have economic interests, that seeks to prevent or delay new demarcations, in order to extract all the possible wealth. In addition, public authorities have been forced to review the demarcations already made.

The non-protection by the public authorities of the rights of Indigenous peoples and non-compliance with the Constitution itself violates the principle of the dignity of the human person and respect for the multifaceted cultural diversity when production, preservation, conservation and recreation are guaranteed by the law of cultural heritage [Beckhausen 2007].

Many of these criticisms of the State's omission towards Indigenous peoples can be found in the report published by the *Indigenous Missionary Council* (CIMI, acronym in Portuguese), an organization linked to the *National Conference of Bishops of Brazil* (CNBB, acronym in Portuguese), which shows Indigenous peoples are increasingly victims of racism, threats, abuse of power or death. The report also draws attention to the fact that in the face of the non-demarcation of Indigenous lands, especially in the Amazon region, developmental projects interested in the resources of the region have grown enormously, with the complicity of the State itself [CIMI 2016].

Protecting Indigenous Life

Indigenous peoples have a long tradition of respect and sustainable use of their ancestral territories. Their connection to the land is based on the principle that people live in a foreign land. The land is borrowed by one generation because it belongs to the children of the present generation and future generations [Tiradentes & Silva 2008].

Hence, the relationship with the land implies that all are possessors of the planet and not owners of it. Thus, Indigenous peoples do not have a concept of ownership *per se* and assume a duty to respect and protect the land, territory and resources for the enjoyment on an equal basis with the other entities that inhabit it. Therefore, the land is not owned by human beings. Humans are a part of it all [Leopold 1949].

This Indigenous belief is not generally well-understood by non-Indigenous people who see this as idealistic. Although sometimes tolerated, these ideas are rarely taken seriously.

The *Declaration on the Rights of Indigenous Peoples* and its contribution to the recognition of Indigenous peoples in their territory touch on the rediscovery of some bioethical principles [UN 2007]. The Declaration, not only clarifies Indigenous people's rights to land, but emphasizes their vision of the future, including future generations, in a way that is different from the State [UN 2007].

All Indigenous peoples demonstrate that there is a commitment to protect nature and the symbiotic relationship between people, other living beings and the earth. This is contrary to an idea that the natural world is less important than economic profit.

The Declaration also shows that it is possible to integrate the Indigenous vision of protecting the land and the richness of cultural diversity with development, sustainability and respect for human rights. It also provides a glimpse of Indigenous development, coupled with global and deep bioethics and reverence for life [Potter 1988; Schweitzer 2009; UN 2007].

Among Indigenous peoples there is a tradition about a common form of collective land ownership. Their close relationship with the land must be recognized and understood as the fundamental basis of their cultures, their spiritual life, their integrity, and their economic survival. For Indigenous communities, the relationship with land is not merely a matter of ownership and production, but material and spiritual.

The UN *Declaration on the Rights of Indigenous Peoples* fundamentally develops the issue of individual and collective rights to the lands, territories and resources of Indigenous peoples, including their spiritual relationship with waters, coastal seas and other resources, such as fauna and flora. Indigenous peoples have a particular connection with land, which is far removed from their economic or speculative value.

It is, therefore, essential for development and cultural survival, that the UN *Declaration for Indigenous Peoples* recognizes this deep spiritual bond with the land, territories and resources in conjunction with the harmonious system of the environment. It is a significant step towards the protection of the planet and all living beings, Indigenous and non-Indigenous.

As a matter of principle, the UN Declaration recognizes the historical injustice suffered by Indigenous peoples due to the alienation of their lands, territories and resources since the time of colonization. This prevented them from exercising their right to development according to their own interests, which is why the UN in this Declaration maintains that the rights intrinsic to them must be respected [UN 2007]. Dignity and respect for Indigenous peoples involved a long transition from a humiliating initial treatment, to the current condition of protection and subjection, where they still struggle to be respected, to take on the value of self-recognition and to demand the recognition of others.

Self-determination is closely linked to territory. Control of their lands means to exercise the right to determine their destiny and respect their autonomy, their internal government, their own way of life and have the means to finance themselves [UN 2007 (Article 7)]. This allows the maintenance and strengthening of institutions that give meaning to their political, economic and cultural organization [UN 2007 (Article 5)]. Therefore, there is a strong relationship between autonomy and the right over territory and the resources that traditionally Indigenous peoples have used or possessed. This right of Indigenous peoples to control their territory allows them to determine their strategies and priorities regarding the use that is made of the territory and its resources. The State is obliged to consult Indigenous representatives before adopting laws or implementing programs that affect their lands, territories or resources, through the rules of prior consultation in order to obtain their free prior informed consent, particularly when this implies the exploitation of their mineral, water or similar resources [UN 2007 (Article 32)].

Ignorance of the ancestral rights of Indigenous communities over their territories affects other basic rights, such as the right to cultural identity and the very survival of Indigenous communities and their members. Since the enjoyment and effective exercise of the right to communal land ownership ensure that members of Indigenous communities preserve their heritage, States must respect this special relationship to ensure their social, cultural and economic survival. Likewise, the close relationship of the territory with its traditions, customs, languages, arts, rituals, knowledge and other aspects of the identity of Indigenous peoples needs to be acknowledged.

Conclusion

In the *Universal Declaration on Bioethics and Human Rights*, 2005, and in the *Declaration on the Rights of Indigenous Peoples*, 2007, we find the main arguments defended herein. Bioethics can and should occupy a prominent place in the promotion and protection of Indigenous peoples. As indicated, one of the noblest goals of bioethics is to be a place for dialogue between different cultures and moralities, respecting the pluralism of positions. However, in the light of what has been said about Brazilian Indigenous populations, the question is: what can they teach us? To whom is the knowledge, produced from this reflection, directed? It should be remembered that the social and political organizations in an Indigenous society are focused on the collective, especially with regard to the production of knowledge.

Indigenous peoples need to be recognized in their diversity, autonomy and protagonism in their relations with the various facets of the national State, thus recognizing this central role should be a precondition in discussions on Indigenous bioethics.

A bioethics of protection, one of the Brazilian themes of bioethics, attempts to address such problems in a critical and impartial manner, seeking to overcome merely anthropocentric view and in so doing, protects us from ourselves. From the beginning, it is clear that there is a common thread in which respect for land and life goes hand in hand with the very existence of Indigenous peoples, both individually and collectively. Therefore, there is a co-responsibility to teach this way of being going forward.

Our discussion shows it is possible to perceive the compatibility of the Indigenous peoples' vision with the search for bioethical principles. Albert Schweitzer, one of the forerunners of bioethics, developed a "reverence of life" ethics, in which he understood that the will to live among others involves the promotion of and a critical revision of our way of life, thus discovering co-responsibility for all living beings [Schweitzer 2009]. This conception of bioethics is not limited to certain areas of knowledge (medical, environmental, technological, philosophical), but implies an expansion of the individual and collective way of seeing the world.

References

Beauchamp, T. L. and Childress, J. F. (2012). *Principles of Biomedical Ethics*, 7th Edition (Oxford University Press, Oxford).

Beckhausen, M. (2007). Direitos indígenas, *Revista Eletrônica* PRPE. Available from: <http://www.prpe.mpf.mp.br/internet/index.php/internet/Revista-Eletronica/Revista-Eletronica/2007-ano-5/Dissertacao-de-Mestrado-sobre-direitos-indigenas>.

Brazilian Federal Constitution. (1988). *Constituição Federativa da República do Brasil.* Available from: <http://www.senado.leg.br/atividade/const/con1988/con1988_14.12.2017/ind.asp>.

CIMI. (2016). *Violência contra os Povos Indígenas no Brasil.* Available from: <https://www.cimi.org.br/pub/relatorio/Relatorio-violencia-contra-povos-indigenas_2016-Cimi.pdf>.

Dantas, F. A. C. (2006). Base jurídica para a proteção dos conhecimentos tradicionais. *Revista* CPC 1(2): 80–95. Available from: <http://www.revistas.usp.br/cpc/article/view/15590/17164>.

Kleinman A. (1978). Concepts and a Model for the Comparison of Medical System as Cultural Systems, *Social Sciences and Medicine.* 12: 85–93.

Leopold, A. (1949). *A Sand County Almanac and Sketches Here and There* (Oxford University Press, Oxford).

Pessini, L. (2016). Bioética: algunas interrogantes que desafian el presente y el futuro de America Latina. Ferrer, J. *et al.* (eds.) *Bioética: el pluralismo de la fundamentación* (Universidad Pontifícia Comillas, Madrid) pp. 419–432.

Pessini, L., Barchifontaine, C.P. and Stepke, F.L. (eds.). (2010). *Ibero-American Bioethics. History and Perspectives* (Springer Netherlands).

Potter, V. R. (1971). *Bioethics. Bridge to the Future* (Prentice-Hall, Englewood Cliffs, New Jersey).

Potter, V. R. (1988). *Global Bioethics. Building on the Leopold Legacy* (Michigan State University Press, East Lansing, Michigan).

Schweitzer, A. (2009). *Out of my Life and Thought. An Autobiography,* 60th anniversary edition (John Hopkins University Press, Baltimore).

Schramm, F. R. (2008). Bioética da proteção: ferramenta válida para enfrentar problemas morais na era da globalização, *Revista Bioética* 16(1): 11–23. Available from: <http://www.redalyc.org/html/3615/361533250002/>.

Schramm, F. R. and Kottow, M. (2001). Principios bioéticos en salud pública: limitaciones y propuestas, *Cadernos de Saúde Pública* 17(4): 949–956. Available from: <https://www.scielosp.org/scielo.php?pid=S0102-311X2001000400029&script=sci_arttext&tlng=>.

Tiradentes, J. A. and Silva, D. R. (2008). *Sociedade em construção: história e cultura indígena brasileira* (Direção Cultural, São Paulo).

United Nations. (1948). *Universal Declaration of Human Rights.* Available from: <http://www.un.org/en/universal-declaration-human-rights/index.html>.

United Nations. (2007). *Declaration on the Rights of Indigenous Peoples.* Available from: <http://www.un.org/esa/socdev/unpfii/documents/DRIPS_en.pdf>.

United Nations Educational, Scientific and Cultural Organization (UNESCO). (2005). *Universal Declaration on Bioethics and Human Rights.* Available from: <http://www.unesco.org/new/en/social-and-human-sciences/themes/bioethics/bioethics-and-human-rights/>.

World Health Organization (WHO). (2013). *WHO Traditional Medicine Strategy, 2014–2023.* Available from: <http://apps.who.int/medicinedocs/documents/s21201en/s21201en.pdf>.

https://doi.org/10.1142/9781786348579_0004

CHAPTER FOUR

Bioethics and Human Rights Issues of Indigenous Peoples in Japan, with Particular Focus on the Ainu

Taketoshi Okita,[*] *Atsushi Asai,*[†]
Masashi Tanaka[‡] *& Yasuhiro Kadooka*[§]

Abstract

Two Indigenous peoples, the Ainu and the Okinawans, reside in Japan. The former live in the northernmost part and the latter live in the southernmost. In this paper, we present the history of and contemporary issues surrounding discrimination against the Ainu, and consider bioethical issues pertinent to them. First, we refer to a long history of opposition toward the Ainu in Japan. The Ainu have suffered serious long-term discrimination and human rights violations, including the deprivation of their ethnic autonomy and traditional residence; murder; forced assimilation, labor, and migration; compulsory withdrawal of children; and

[*] Associate Professor, Department of Medical Ethics, Tohoku University Graduate School of Medicine, Sendai, Miyagi, Japan.
[†] MD, PhD, MBioeth., Professor, Department of Medical Ethics, Tohoku University Graduate School of Medicine, Sendai, Miyagi, Japan.
[‡] MD, Department of Medical Ethics, Tohoku University Graduate School of Medicine, Sendai, Miyagi, Japan.
[§] Professor, Department of Bioethics, Kumamoto University Faculty of Life Sciences, Kumamoto, Kumamoto, Japan.

destruction and loss of their language, culture, religion, traditions, and customs. Second, we consider research using the Ainu in Japan. In the name of science, including phrenology, extensive excavations of Ainu's human remains had occurred without informing living Ainu residents or obtaining their permission. The Ainu people were completely objectified in these research activities. Currently, over 1600 Ainu ancestral remains are stored in twelve Japanese universities. Finally, we consider how we might prevent majority ethnic groups from unethical behaviors and attitudes toward Indigenous peoples, including future research studies that may discriminate and objectify other human beings.

Introduction

Honda [2011] lists four attributes of Indigenous people: indigeneity, being ruled, a continuous and shared history, and general self-awareness as Indigenous people. He explains this definition by stating that Indigenous people and their descendants are those who have lived in a region long before a modern state dominates the area as part of the process of colonization. Under the colonial rulers, they are deprived of their original lifestyle and are placed in inferior social and legal situations, but keep their historical continuity by residing together, and recognize themselves as Indigenous or native people [Honda 2011]. Uemura [2001, p. 281] defines Indigenous people as ethnic groups created as a consequence of modern state colonization policies who are regarded as barbaric and uncivilized; they are unilaterally deprived of their land which is annexed, and their identities and cultures are denied under cultural assimilation policy.

When considering the problems surrounding Indigenous peoples, both in Japan and throughout the world, we must address difficult, serious, and unavoidable questions of how to stop the current repression, discrimination, atrocities, harassment, exploitation, and objectification leveraged against vulnerable people in different ethnic groups. Therefore, while keeping in mind the question of how we might coexist in a single society without exploiting, destroying, or causing suffering among minority peoples, this chapter presents several topics concerning Indigenous people in Japan, with a particular focus on the Ainu.

In the following section (Section 2), we introduce the two known Indigenous in Japan: the Ainu and the Okinawans. Thereafter, we will primarily focus on the Ainu. In Sections 3 and 4, we will discuss the history of and contemporary issues surrounding discrimination against the Ainu in our country. Section 5 presents some historical cases of bioethics and human rights issues related to medical research on the Ainu people, as well as some potential future perspectives. We conclude by proposing what we should or can do to prevent unethical acts and abuses of human rights against Indigenous peoples in the future.

Indigenous People in Japan

The Ainu and the Okinawans have been recognized as Indigenous people in Japan, with the former living on the northernmost island (Hokkaido) and the latter on the southernmost island (Okinawa) of the country [Hansen et al. 2017]. With regard to the development of ethnic groups in Japan, Hanihara argues: "The first occupants of the Japanese Archipelago came from somewhere in Southeast Asia in the Upper Palaeolithic age and gave rise to the people in the Neolithic Jomon age, or Jomonese; then the second wave of migration from North Asia took place in and after the Aeneolithic Yayoi age; and the populations of both lineages gradually mixed with each other" [Haniwara 1993]. Hanihara concludes that the Ainu and Okinawans are direct descendants of residents of the Japanese archipelago in the Paleolithic Age [Haniwara 1993].

According to Segawa, the Ainu people have dramatically maintained the characteristics of the Jomon people in the Japanese archipelago. Segawa argues that the majority of today's Japanese population comprises those who came from the Korean Peninsula during the Yayoi era and the Jomon people, the latter of whom are Indigenous peoples of the Japanese archipelago [Segawa 2015]. He also argues that the Ainu are the "original" people of the Japanese archipelago [Segawa 2015]. The early modern era of the Ainu (Ainu culture period) and their collision with the majority Japanese people ("Wajin" or the Yamato ethnic group) [Omoto 2014].

In his article of June 6, 2008 Kato [2017] reports that a bipartisan, non-binding resolution was approved by the Japanese Diet, which called upon the government to recognize the Ainu as Indigenous to Japan; that very same day, the Japanese government finally complied and recognized the Ainu as such. A living condition survey in 2013 (Department of Environment and Lifestyle, Hokkaido Government 2015) estimated that the Ainu population in Hokkaido comprises only 16,786, but this number does not include Ainu populations around Tokyo and Osaka on the main island of Honshu [Kato 2017]. Meanwhile, the Okinawans have yet to be recognized by the Japanese government as Indigenous people [Hansen *et al.* 2017].

Before shifting the spotlight to the Ainu, it is worth noting that Okinawa's Indigenous ethnic group was invaded and dominated by the Satsuma clan of the Edo shogunate in 1609. In 1872, the Ryukyu clan was established in Okinawa by the Meiji Government of Japan and the Yamato Japanese (hereafter, the Yamato) began to discriminate and repress the Okinawans. During this time, the Okinawans were forced to assimilate with the majority Yamato [Nakamura 2017a].

There is also a report from 1903 that states that two Okinawan women made spectacles of themselves as "ladies of the Ryukyus" when they wore ethnic costumes in the hut known as "Humanity Hall" in Osaka on the mainland. At that time, 32 people including the Ainu, Taiwanese Aborigines, Javanese, Indian, and Hawaiian, wore their national costumes and presented their daily living activities to an audience. This human race building was created by Shosuke Tsuboi, a natural anthropologist at the Tokyo Imperial University at the time. As is well known globally, Okinawa is currently facing serious political issues pertaining to the relocation of the U.S. military base [Nakamura, 2017a]. Although the present paper does not address ethical, legal, and social issues concerning the Okinawans, we believe that they share common serious problems facing other Indigenous peoples, including the Ainu people of Hokkaido.

A Long History of Opposition Towards the Ainu in Japan (prior to World War II)

The following historical summary is based on Takuro Segawa's "Introduction to Ainu Studies" [2015] and the history of the Ainu people in the Japanese archipelago described by Keiichi Omoto [2014]. From the 13th to the 16th century, the Ainu people in Hokkaido conducted free trade in the Kuril Islands, Honshu, and Mainland China. The Yamato of the main island (Honshu) acquired products from the north, which brought great wealth to them through the Ainu. However, when the trade expanded, the Yamato emigrated northward to Hokkaido beginning in the latter half of the 14th century, in order to monopolize its wealth. Starting with Koshamain's battle that occurred in Donan in 1457 between the Ainu and the Yamato, the Ainu uprising continued for a century afterward [Segawa 2015].

The battle of Koshamain (1457) was waged because of the murder of a young Ainu man by the Yamato. The Yamato had suppressed the Ainu, and even after several decades there was intermittent suppression of the Ainu uprising by the Yamato [Omoto 2014]. The lord of the Matsumae clan in 1550 brought the fight to an end. The Matsumae clan squeezed the life out of the Ainu with a system that benefited the Yamato, and the battle of Henaouque (1643) occurred right after the turbulence of the Shimabara/ Amakusa (1637–38) in Kyushu. Despite many unknown details, some Ainu people allegedly were dissatisfied with highly unequal conditions imposed upon them by the Matsumae clan and attacked the Yamato. This was the first war following the establishment of the Matsumae clan [Omoto 2014].

The battle of Shakshin, the biggest Ainu war in the Edo period, took place in 1669. The attack by Shakshin, the Ainu's chief, was triggered by a misdemeanor committed by the Matsumae clan, as well as the forced entrainment of the Ainu as a labor force, the destruction of the Ainu villages (Kotan), the conflict between the Ainu and the Yamato, and the poisoning of the Ainu people by the Yamato. Shakshin called for the simultaneous uprising of other Ainu people who had been dissatisfied with their circumstances and treatments imposed on them in different Kotans and they attacked 19 Japanese merchant ships and killed 273 Yamato people. The Matsumae clan cheated Shakshin and poisoned him to death, and the clan suppressed the Ainu's uprising two years later. As a result, the Matsumae clan's controls on the Ainu became stronger and more numerous [Omoto 2014].

During the 18th century, the Ainu were used as cheap labor and were made to work under harsh conditions. They were forcibly taken from their villages and moved to specific fishing areas, and the traditional Ainu society collapsed. The Yamato despised the Ainu, calling them foolish. The Yamato deceived them, looted their fortune and wielded violence against them. Due to the occupation by the Yamato people, various diseases spread among the Ainu, leading to drastic population decreases. The battle of Kunashiri-Menashi (1789) was the last effort of organized resistance by the Ainu and was brought upon by the abuse against them by the Yamato. By that point, the Ainu had been long enslaved and were no longer able to endure the tyranny. The Ainu reportedly killed 70 Yamato in this battle, and lost 37 of their own [Segawa 2015; Omoto 2014].

According to Omoto, the Edo shogunate adhered to the use of the name, "Ezo" to describe the land of Hokkaido. Since ancient times, Hokkaido was called Ezo, which encompassed the meaning of "savage people," and indicated "the land where the unexplored people live." After the Matsumae clan advanced, it was decided that the name "Ezo" was to be used to pinpoint specifically the Ainu's place of residence [Omoto 2014]. According to Kato, in 1868, after the Meiji Restoration, Hokkaido was officially annexed by Japan. The Ainu language was banned, and the Ainu were forced to take Japanese names. Then, under the assimilation policies such as the Hokkaido Former Aborigine (Dojin) Act of 1899, the Ainu were prohibited from performing traditional activities that were necessary for their life and culture. This Act was enforced for 98 years until 1997 [Kato 2017].

Through the Hokkaido Former Aborigine Protection Act (1899–1997), the Meiji government forced the Ainu to "Japanize" in all aspects. Funeral ceremonies were no exception [Hirata 2016]. Notably, the term "*Dojin* (aborigine)" in Japanese contains deep contempt and means Indigenous people who are uncivilized. Evidently, a perception of racial superiority and discrimination against the Ainu by the Yamato has existed for a very long time.

According to Momose [1994], the problem with the Hokkaido Former Aborigine Protection Act is that, while it had actually provided a means to promote agriculture and assimilation of the Ainu, thus offering support and protection for them, it was not properly enforced. The administration framed the Ainu as a group needing national protection, perpetually placing the Ainu in the disadvantageous situation as an aborigine (*Dojin*) [Momose 1994]. In 1997, the "Act on Promotion of Ainu Culture and the

Dissemination and Enlightenment of Knowledge on Ainu Traditions" was promulgated and enforced; at the same time the Hokkaido Former Aborigine Protection Act was abolished [Momose 1994].

Shigeru Kayano was born into the Ainu ethnic group, and he later became a member of the House of Councilors who was active in the Diet. He published his autobiographical memoir, entitled *Ainu no Hi (Monument of the Ainu),* comprising several chapters on topics, such as "My Grandfather was a slave of the Japanese (the Yamato)," "The end of compulsory immigration," and "My father made a sinner" [Kayano 1980]. Below, we summarize his experiences, observations, and opinions in the book concerning several relevant issues.

With regard to the "Hokkaido Former Aborigine Protection Act," Kayano [1980] wrote that the Ainu never were *"Dojin* (aborigine)"; rather, they were citizens who had been living in Ainu Mosiri (human quiet earth), in Hokkaido. The Yamato invaded this land. Ainu Mosiri was a territory unique to the Ainu people. The Yamato invasion began in the Meiji Era; the Hokkaido Former Aborigine Protection Act ignored the Ainu's fundamental right to live as hunters, and instead gave them poor land where they were forced to become farmers. This robbed the Ainu of their freedom, and wrongly justified the Yamato's deprivation of the land from the Ainu, representing a complete invasion [Kayano 1980].

Kayano also describes the forced and cruel exploitation of the Ainu by the Yamato: When a child reached a certain age, the Yamato would forcibly take him or her from their home and make them work. The remuneration was quite insignificant and sometimes nonexistent. Kayano's grandfather was taken at the age of twelve, along with roughly half of his village's population. His grandfather was forced to march 350 km from Nibutani to the worksite destination. He was not given a place to sleep at the worksite and received little to eat. Many of the younger children became ill or died of exposure to cold and hunger [Kayano 1980].

Kayano's recollection of the forced migration is that in the Meiji era, the Yamato invaded the Ainu land and began forcing the Ainu to emigrate to other unfavorable places. The Ainu contracted tuberculosis and died in vast numbers due to the hard labor conditions combined with the food shortage [Kayano 1980].

History of Discrimination against the Ainu: Cases that Occurred After World War II

Many incidents have arisen due to discrimination against the Ainu people after World War II. We review these here and refer to some relevant cases and information therein.

The Hashine case

In October 1972, Naohiko Hashine, a day laborer in Tokyo, got into a fight at a drinking party with Sho Shu Kim, a Korean living in Japan. Hashine stabbed Kim to death with a fruit knife after Kim made a pejorative reference to Hashine's identity as an Ainu. Hashine was sentenced to prison for five years. It is worth noting that this incident occurred between two individuals who were both discriminated against by Japanese society [Winchester 2014]. Masao Sasaki wrote, "What if it had been I who was abused as this Ainu was? There have been various cases in my past that have on occasion provoked fierce hatred, and sometimes internal intentions for murder. Regardless of how my hands or feet responded, my mind conjured up scenes of killing my opponent at the time. As a person who has had repeated murderous intentions in my heart, I must say that I am no different from Naohiko Hashine" [Sasaki 1974].

The Hokkaido University discriminatory lecture case

Below we present a brief summary of a particular case which was extensively described and analyzed by Tetsuya Ueki in one of his texts concerning the Ainu [Ueki, n.d.]. In April 1977, a faculty member of the Economics Department at Hokkaido University presented the first lecture on the history of Hokkaido economics to 100 students. In this lecture, the professor reportedly said, "Hokkaido's economic history is the history of development mainly driven by *Nihonjin* (the Yamato), and I would omit and ignore the history of the Ainu." He was the Director of the Economics Department at the time. He repeatedly brought up the physical characteristics of the Ainu people, presented derogatory expressions toward Ainu women, and spoke other discriminatory words as a "joke" to make the students laugh. University students and Yuuki Shoji of the Ainu Liberation Union, who consulted with the students, demanded that the professor state his reasons for the discriminatory remarks, but no response was obtained. The students then locked the professor in the classroom, and the police were called to handle the case. The professor was finally rescued by 120 riot police after eight hours of confinement. Three students were arrested for illegal detention. The Association of the Faculty of Economics who refused aggressive intervention was also charged. The arrested students were not prosecuted, but the professor resigned as a Dean. After that, several discussions were held between Yuuki Shoji and the professor, during which the professor officially apologized for his discriminatory remarks in his lecture [Ueki, n.d.].

The Ainu portrait trial

Mikako Chikkapu found out that her photograph had been posted without her permission in a book entitled, *"The Ainu Ethnic Magazine,"* edited by the Ainu Culture Preservation Council and issued by Daiichi Gakkai Publishing Co., Ltd. in 1969. In 1985, Chikkapu filed a lawsuit against the publisher and editorial officials of the book due to the violation of her portrait rights. Apart from the portrait rights, the plaintiff also sued the defendants for writing this book with the working premise that the Ainu people were "an extinguishing ethnic group," and that the Ainu people had been treated like specimens in the book [Kanazawa 2011]. The defendant (editorial officer) acknowledged that he had written a paper for The *Ainu Ethnic Magazine* with the assumption that the Ainu has already disappeared as an ethnic group or race. He also admitted that he had believed that it was good for the Ainu to rid themselves of the traditional Ainu lifestyles, throwing out "bad customs" as a result of the introduction of the Yamato culture. However, at the same time, he also claimed that he did not intend to insult or despise the Ainu ethnic group. It has been argued that it was this trial that revealed "unconscious discrimination" by the Yamato against the Ainu ethnic group. In 1988, the defendant apologized in full and a settlement was reached, which was key to launching the movement to restore the Ainu [Kanazawa 2011].

The Nibutani dam trial

In 1993, a dam was built in the Nibutani district of Hiratori town in Hidaka, Hokkaido, despite opposition from the Ainu people. Representatives of the Ainu people appealed this decision to the Sapporo District court as unfair because the government had illegally seized their land. In 1997, the Court confirmed that construction of the dam was indeed illegal, and the plaintiffs won the suit. It was acknowledged that the Ainu had the right to enjoy their own culture and that, in building the dam, the government was not sufficiently considerate of this right, as they would have destroyed facilities considered important for their ethnic culture. For the first time as a state institution, the court also recognized that the Ainu living in Hokkaido were Indigenous people who had established a unique culture there long before the arrival of the Yamato [Omoto 2014; Kato 2017; Hirata 2016]. This decision suggested that the Ainu had rights that required consideration of Article 13 of Japan's Constitution, which protects the rights of the individual and the International Covenant on Civil and Political Rights [Omoto, 2014; Kato, 2017; Hirata, 2016].

Current Situation Surrounding Discrimination against the Ainu in Japan

Even today, serious problems concerning discrimination and prejudice against the Ainu people remain [Kanazawa 2011]. A report in *The Indigenous World 2017* suggests that public awareness of the Ainu's situation remains problematic. According to a survey conducted by the Japanese government in 2016, 72.1% of Ainu agreed with the statement that "discrimination against the Ainu people exists." Among the general public, however, only 17.9% agreed with the statement, with 50.7% saying that "discrimination doesn't exist" [Hansen *et al.* 2017].

Sasaki reported the details of discrimination against the Ainu in her paper published in 2016. These findings can be summarized as follows: Some typical discriminatory actions against the Ainu include both ridiculing the hairiness of Ainu individuals and saying "Ah, inu ga kita (Oh, here comes a dog!)" in which "Ah, inu (dog)" is used in a play on the word "Ainu". The seriousness of the discrimination depends on the region where the Ainu live, the historical background of communities, and population ratio of the Ainu in an area in question. Perceptions of prejudice and discrimination among the Ainu people tend to decrease with larger population ratios of the Ainu. This discrimination is also weaker among younger generations but is stronger against Ainu women than men. Ainu people tend to experience discrimination as early as elementary school [Sasaki 2016].

Some Ainu people were rejected for marriage by a potential mate simply because they are Ainu. Discriminatory experiences in the classroom during early life stages lower the desire among the Ainu to continue with higher education, which deprives them of the chance to pursue an academic career and leaves them with a poor educational background. Even teachers have discriminated against them. The high school entrance rate of the Ainu is half the rate of the Yamato [Sasaki 2016]. Discriminatory experiences at school have a tremendous influence on their life choices and deprive them of many valuable opportunities. Their ethnicity is a disadvantage for employment as well, and they will often hide their ethnic origin [Sasaki 2016; Kayama 2017]. Some fear that they may be killed if their ethnicity is revealed to the Japanese society, and some suffer from depression due to ethnic discrimination [Kayama 2017].

Segawa published an interview with an Ainu woman (A-san), entitled "Modern Times: Living as an Ainu" [Segawa 2015]. A-san is often asked by the Yamato questions such as, "Can the Ainu people speak Japanese?" and "Are they still living in thatched houses (*Chise*)?" The interview reveals a lack of adequate understanding of contemporary Ainu's life among the Yamato. It also reveals the remaining prejudice that Ainu people are dumb [Segawa 2015].

Scientific Research Involving the Ainu and the Objectification of Research Subjects

Massive excavation of the remains from the Ainu ethnic group

Phrenology became popular in the 19th century in Europe and the United States, and skull collections were common among researchers. At the time it was thought that the characteristics and superiority/inferiority of "race" could be clarified and demonstrated by measuring the shape and size of the skull, and that the skull volume reflected the size of the brain, which in turn reflected the mental ability and the level of intelligence. Phrenology was believed to be able to prove the spiritual superiority of Caucasian individuals over black and Indigenous peoples [Ueki 2017]. Meanwhile, the collection of Ainu skulls in Japan began for the pursuit of origins among the Japanese. As a part of the national project, entitled "Medical and ethno-biological investigational research concerning the Ainu" (1931), Kodama and other researchers started collecting the remains of the Ainu from their cemeteries in Hokkaido and other areas [Kato 2017; Ueki 2017]. It has been pointed out that since its establishment in 1921, the School of Medicine, Hokkaido Imperial University (present-day Hokkaido University), has had strong research interests in the Ainu people [Nakamura 2017b].

Yoshinori Koganei, who conducted massive excavations in Ainu cemeteries in 1888, was reported to say, "It is most important to avoid places where the Ainu people still reside and instead seek an old, unrelated tomb." He collected Ainu remains without informing the living Ainu residents or obtaining their permission to perform his excavations. Kenji Seino also excavated the skulls of the Sakhalin Ainu during the Taisho Period. Both Koganei and Seino chose not to excavate the remains in front of the Ainu. They were conscious of the fact that cemetery excavation was a crime against the Ainu and had been paying enough attention not to be found out [Ueki 2007].

In the 1930s, Sakuzaemon Kodama, a professor of anatomy in the School of Medicine at Hokkaido University amassed a huge collection comprising more than 1,000 Ainu skulls. Although this was a major contribution to research on the Ainu, his reputation has been seriously tarnished by accusations of grave robbery and artifact theft [Scott 2013]. The excavations by Kodama have been criticized by the Ainu people [Ueki 2005]. Kodama also collected buried objects including swords, hunting materials, earrings, and pans [Hokkaido University 2013]. Currently, over 1600 Ainu ancestral remains are stored in 12 Japanese universities [Kato 2017].

Criticism of Tetsuya Ueki on the excavation of Ainu remains

Kodama justified his criminal excavation of the various remains from cemeteries by regarding the cemeteries as ruins. However, his collection contained bones that were buried only several decades ago. Many Yamato educational officials and those with administrative power also cooperated with his actions. Kodama advocated for the urgency of excavation and emphasized the significance of academics [Ueki 2007; 2017]. Since the Ainu represent an "extinguishing ethnic group," he argued, the culture, language, and physical dispositions must be examined before their "extinction" and that records must be kept. According to Ueki, this logic has been frequently used by researchers of the Ainu in the same generation [Ueki 2007; 2017].

Ueki also argues that academic conservation of the Ainu culture and ethnicity was regarded as the responsibility of scholars, and that promotion of this research was considered appropriate and more important than avoiding inhumane acts against the Ainu. The authority of research felt in society as a whole supported such researchers' rights. Japanese society at the time had accepted these researcher claims, the excavation had been praised by society, and no ethical concerns were acknowledged. That said, there was strong opposition toward the excavation from the Ainu side, as revealed in many testimonials that reviewed the excavations and criticized them uniformly and severely [Ueki 2007; 2017].

In his article in *The Japan Times,* Simon Scott quotes words from the aforementioned author, Kayano, in his text "*Ainu no Hi (*English title, *Our Land Was A Forest: An Ainu Memoir)*," with regard to the academic investigations on the Ainu conducted by Japanese researchers [Scott 2013; Shigeru 1994].

In those days I despised scholars of Ainu culture from the bottom of my heart…Each time they came to Nibutani, they left with fork utensils. They dug up our sacred tombs and carried away our ancestral bones…Under the pretext of research, they took blood from villagers and, in order to examine how hairy we were, rolled up our sleeves, then lowered our collars to check our backs [Shigeru 1994].

Scott writes that particularly painful memories for Kayano were those of his mother subjecting herself to this so-called "scholarship," after which Kayano recalled her staggering home, weakened from having her blood taken for research purposes. What seemed to disturb Kayano the most, though, more than the pillaging of cultural artifacts and human remains, was the dehumanizing way this research was carried out [Scott 2013; Shigeru 1994].

> Kayano writes: There was also portrait photography. People not only were photographed from the front, the side, and an assortment of angles but induced to wear large number plates such as criminals wear in mug shots. Among the photos of my mother is one in which a number plate hangs from her neck [Shigeru 1994].

We would argue that the Ainu people were completely objectified in the aforementioned research activities concerning Ainu remains. Typically, objectification involves treating persons or other non-objects as things [Eguchi 2006]. According to the definition of Orehek and Weaverling [2017], people are objectified when they are treated as a means to an end, used in the same way as objects, or evaluated according to their instrumentality to the goals of others. We think that Yamato researchers stole the remains of the Ainu people, processed and measured them as research materials, converted them into samples or exhibits, and kept them without consideration for anything other than their own interests or purposes. We must conclude that consideration and respect for the Ainu were severely lacking.

Current status of the return of the Ainu remains and future research activities

In 2013, the Chief Cabinet Secretary decided to set up a new national museum for the Ainu people as well as a memorial facility to store unidentified Ainu ancestral remains, which are presently stored in universities across Japan, in Shiraoi, Hokkaido (scheduled to open in 2020) [Kato 2017]. According to the report by Ueki in 2017 [Ueki 2017], the current status of the return of Ainu remains is that only 35 bodies have been returned. There are 1,636 sets of bones at 12 universities nationwide, and only 23 sets of bones that can identify individuals (Hokkaido University 19, Sapporo Medical University 4).

The Ainu volunteers of the Hidaka district formed the "Kotan no Kai" (the society of traditional Ainu communities), and this group received court approval to become the recipient for the return of these remains [Ueki 2017]. A group of five Ainu had initiated court proceedings against the university for the return of the remains. In March 2016, Hokkaido University agreed, under a court-mediated settlement, to return the remains of 16 Ainu people to the Ainu of Urakawa, Hokkaido, which had been excavated 85 years ago for research purposes [Hansen *et al.* 2017].

There is also an ongoing debate concerning whether or not the afore-mentioned research has benefited the Ainu people [Ueki 2007; 2017]. Japanese researchers involved in academic studies on Ainu remains have argued that past anthropological or medical research for the Ainu has actu-ally served their interests, while acknowledging the unethical nature of the past research activities. For example, Dodo [2015] argues that past anthro-pological research using Ainu remains seems to have benefited them when balancing the benefits gained and losses inflicted on the them. Dodo rea-sons that these research studies based on the excavation of cemeteries and measurement of the skulls in the past revealed that the Ainu represent an Indigenous people of the Japanese archipelago — this finding is undoubt-edly something considered profitable for the Ainu people.

On the other hand, Hatozawa, an Ainu commentator, called the researchers specializing in the Ainu "vermin", because he felt that these scholars were trying to raise the value and significance of their research by emphasizing the possibility of Ainu extinction, while remaining bystanders, doing nothing to stop the decline of the Ainu ethnic group [Hatozawa 1972]. Tsunemoto, a folklorist in Japan, pointed out the many "plunder investigations" in which researchers never returned borrowed materials from research participants or cooperators nationwide. He also reveal some researchers' arrogant attitudes towards research subjects and overestimation of the importance of their academic activities [Miyamoto 2008].

Conclusion

Our review of a wide variety of materials and views of both the Ainu people and researchers in the field of Ainu history and relevant academic research has revealed and demonstrated that the Ainu have suffered serious long-term discrimination and human rights violations. These have included the deprivation of their ethnic autonomy and traditional residence; murder; forced assimilation, labor, and migration; compulsory withdrawal of children; and destruction and loss of their language, culture, religion, traditions, and customs. They have been subjected to persistent and massive plunder of their ancestors' remains and ceremonial supplies, in the name of research. We would also assert that the Ainu have long been exploited and objectified by the Yamato, and they continue to be exposed to bullying, poverty, and hate speech, and receive inadequate education opportunities in present Japan. Their real names have been lost and they have been deprived of individual and community identity.

All of this discrimination toward the Ainu described above was brought about by the Yamato ethnic group to which we modern-day Japanese belong. The Yamato have long ignored the dignity of the Ainu as human beings and persons, exploited them for our own gain, and denied their very existence. In the past, political claims that Japan is a single ethnic nation were widespread and accepted, and it was commonly believed that "Japanese" was synonymous with "Yamato" [Segawa 2015; Kayama 2017]. In fact, we authors were taught this when we were children, and have accepted this assumption for decades. Perhaps this is the truth for the majority of Japanese individuals with no Ainu heritage.

In 2016, another serious case of discrimination by the modern Japanese (Yamato) toward Indigenous people in Okinawa was reported [Kayama 2017; Sato et al. 2016]. Two male riot police in their twenties, dispatched from the Osaka Prefectural Police to provide security for the relocation of the US Army helipad in Okinawa, said to some of the Okinawans protesting against the relocation of the US base, "You, Dojin (Aborigine)" [Kayama 2017; Sato et al. 2016]. Okinawa Governor Takeshi Onaga spoke on January 19, stating that Dojin is a word implying strong contempt against local citizens, and therefore entirely unacceptable and outrageous. The Minister Chief Cabinet Secretary Suga also pointed out at a press conference that these improper remarks made by the two policemen were highly regrettable and unforgivable [Sato et al. 2016]. Indeed, this is a very disappointing case that suggests that the Yamato discrimination against Indigenous peoples in Japan has been unfortunately conveyed to the younger generations of our country.

Discrimination is the poor treatment of specific individuals or groups for unjustifiable reasons. According to Kayama, a psychiatrist, human beings tend to attack "differences" in others and feel or think that those who differ from us are inferior. Kayama also points out that acts of discrimination, bullying, and harassment all attack these differences [Kayama 2017]. We would also argue that the thoughts and logic based on sham science such as eugenics wrongly justify discrimination, objectification of any individuals, and human rights violations. Furthermore, we would caution that the involvement of the interests of the state and one's own ethnic groups opens the door wide to blind and inhumane acts such as invasion, control, repression, and exploitation of Indigenous peoples.

In a similar vein, eugenic ideas and attitudes of academic supremacy would spur researchers to act condescendingly and objectify their research collaborators and participants as well. We would argue that no research study is completely free of cultural environments and historical constraints, and thus it is impossible for anyone to conduct research independently from their cultural background and economic interests [Ueki 2017]. Harper [2000] claims that researchers are the products of their times, and that if the intellectual and cultural traditions from which they emerged were permeated thoroughly by insidious bigotry, then the researchers are likely to have eugenic and racist prejudices, no less than any other average human being.

In the final section of this chapter, we would like to consider how we might prevent majority ethnic groups from unethical behaviors and attitudes toward Indigenous peoples, including future research studies that may discriminate and objectify other human beings. Our ideas are as follows:

1. We must face the history of our evil deeds toward Indigenous people and teach these facts to future generations from a neutral and reasonable position. It is important that history is not forged in a manner that would distort it conveniently for our own sake. We should not deny the existence of our inconvenient history and should not hide it. The

direct testimony of stakeholders is important [Hare 2005], and accurate education and reporting of history are essential.

2. We cannot be completely free from our own prejudice. It is important that we are very conscious of the distorted and limited recognition we possess due to our own prejudice and self-interests. If we are researchers, then there is always the possibility of assuming academic superiority over our research collaborators.

3. We cannot emphasize enough the importance of human rights and ethics education. As one author wrote elsewhere, discrimination must be eliminated in order to stop the harm it causes to targeted groups and its impact on their biological and social lives. The only chance for full realization of human rights and ethics is given through a fight against weakness, arrogance, selfishness, self-centeredness, narrow-sightedness, and the failure to recognize and appreciate the suffering of others. This will likely be a long battle [Asai *et al.* 2014].

4. We should not relate differences among people to their superiority/inferiority and we must be aware of our own tendency to attack "differences," as well as the habit of feeling and thinking that people who are different are inferior to ourselves. We must consciously resist our own disposition. Discriminatory ideas and feelings reside in the dark side of the human spirit, regardless of how they are developed or established. Because discriminatory thoughts and feelings are thought to comprise part of the basic human mentality [Akamatsu 2005], it is essential to look human darkness in the face in order to begin to conquer discrimination [Asai *et al.* 2014]. Our attention should be turned to shared sadness, suffering, and joy of human existence, rather than excessive investment in examining our differences.

5. Finally, it is important for all of us to practice empathy by putting ourselves in another's shoes. While this claim is lacking in novelty, it has yet to be fully materialized. Education that teaches the importance of empathy is essential and will always be needed. Despite the dark tendencies mentioned above, humans most certainly can show empathy toward others, as we possess a natural compassion that renders us unable to remain indifferent in the face of another's misfortune [Jullien 2002]. According to Akamatsu, a folklorist studying various types of discrimination in Japanese communities, "The animal called

the human being is hopeless. But we cannot afford to give up antidis-crimination activities. There is no choice but to continue the activities with patience" [Akamatsu 2005].

References

Akamatsu, K. (2005). *Sabetsu no Minzokugaku (Folk Studies in Discrimination)* (Chikuma Gakugei, Bunko, Tokyo).

Asai, A., Ishimoto, H. and Masaki, S. (2014). An Ethical Review of the Production of Human Skeleton Models from Autopsied Patients with Hansen's Disease in Pre-war Japan, *Ritsumeikan Journal of Asia Pacific Studies* 33: pp. 153–161.

Dodo, Y. (2015). Ainu to Jomonjin no Kotsugakuteki Kenkyu: Hone to Katariatta 40nen (Ainu and Jomon Population History: Reflections from a Lifetime of Osteological Research) (Tohoku University Press, Sendai).

Eguchi, S. (2006). Seiteki Monoka to Sei no Rinrigaku (Sexual Objectification and Ethics of Sex), *Contemporary Society, Faculty for the Study of Contemporary Society, Kyoto Women's University* 9: pp. 135–150.

Haniwara, K. (1993). Nihonjin no ru-tsu (The Roots of Japanese people), *Japanese Journal of Geriatrics* 30: pp. 923–931.

Hansen, K. B., Jepsen, K. and Jacqueline, P. L. (Eds.). (2017). *The Indigenous World 2017*. (International Work Group for Indigenous Affairs, Denmark) pp. 304–310. Available from: <https://www.iwgia.org/images/documents/indigenous-world/indigenous-world-2017.pdf>.

Hare, D. (2005). Preface. In D.E. Lipstadt (ed.) *Denial: Holocaust History on Trial* (Harper Collins Publishers, NY) pp. 1–6. Japanese translation by Yayoi Yamamoto (2017) *Hitei to Koutei: Horokousuto no Shinjitsu wo meguru Tatakai* (Harper Books).

Harper, K. (2000). Give Me My Father's Body: The Life of Minik, The New York Eskimo (Profile Books, London).

Hatozawa, S. (1972). Wakaki Ainu no Tamashii: Hatozawa Samio Ikousyu (Young Ainu's Soul: Hatozawa Samio Manuscript Collection). Shin Jinbutsu Ouraisya, Tokyo 35.

Hirata, T. (2016). Feature articles: Ainu. *Syukan Kinyobi* 1101: pp. 22–31 (in Japanese).

Hokkaido University. (2013). Hokkaido Daigaku Igakubu Ainu Jinkotsu Syuzoukeii ni kansuru Tyousa Houkokusyo (Report on the Ainu Human

Remains Stored in the School of Medicine, Hokkaido University) (Hokkaido University, Sapporo).

Honda, T. (2011). Sekai no Senjuminzoku: Nihon no Senjuminzoku (Indigenous Peoples of the World: Indigenous Peoples of Japan), *Gakujutsu no Doukou*, 16(9): pp. 66–69. Available from: <https://www.jstage.jst.go.jp/article/tits/16/9/16_9_9_66/_pdf>.

Jullien, F. (2002). *Doutoku wo Kisozukeru (Foundation of Morality)* (translated by T. Nakajima and Y. Shino). (Kodansha Gendai Shinsho, Tokyo).

Kanazawa, E. (2011). Chikappu Mieko Shi to Ainu Minzoku Undo: Hyoron 2010nen no Hokkaido (Chikappu Mieko and Ainu Ethnic Movement: Criticism Hokkaido in 2010), *Sapporo University Research Institute Journal* 2: pp. 273–277.

Kato, H. (2017). The Ainu and Japanese Archaeology: A Change of Perspective. *Japanese Journal of Archaeology* 4: pp. 185–190.

Kayama, R. (2017). 'Ijime' ya 'Sabetsu' wo Nakusutame ni Dekiru Koto (What We Can Do to Eliminate 'Bullying' and 'Discrimination') (Chikuma Primer Shinsho, Tokyo).

Kayano, S. (1980). *Ainu no Hi (The Monument of Ainu)* (Asahi Shinbunsya, Tokyo).

Miyamoto, T. (2008). Chousachi Higai: Sarerugawa no Samazamana Meiwaku (Survey Site Damage: Various Annoyance on the Side Being Surveyed). In Ankei, Y. and Miyamoto T. (eds.) *Chousasareru to iu Meiwaku: Field ni deru mae ni Yondeoku Hon (An Annoyance to be Investigated: A Book to Read Before Going into the Field)* (Mizunowa Shuppan, Suou Oshima) pp. 13–34).

Momose, H. (1994). Hokkaido Kyu-Dojinhogohou no Seiritsu to Hensen no Gaiyou (The Enactment and Revision of the Protection Act for the Ainus in Hokkaido of 1899: A Historical Outline), *Shien* 55(1): pp. 64-86.

Nakamura, K. (2017a). Honne de Kataru Okinawashi (Truth Talk about the History of Okinawa) (Shintyou Bunko, Tokyo).

Nakamura, N. (2017b). Cultural Affiliation is not Enough: The Repatriation of Ainu Human Remains, *Polar Record* 53: pp. 220–224.

Omoto, K. (2014). Senjuminzoku to Jinken (1): Ainu to Senju Amerika jin (Indigenous People and Human Rights (1): Ainu and Native Americans). *St. Andrew's University Bulletin of the Research Institute* 29(3): pp. 101–120.

Orehek, E. and Weaverling, C.G. (2017). On the Nature of Objectification: Implications of Considering People as Means to Goals, *Perspectives on Psychological Science* 12(5): pp. 719–730.

Sasaki, C. (2016). Dai 3 shou Gendai niokeru Ainu Sabetsu (Chapter 3 Ainu Discrimination in Modern), *'Chousa to Syakai Riron' Kenkyu Houkokusyo ('Survey and Social Theory' Research Reports)* 35: pp. 45–70.

Sasaki, M. (1974). Ima, Syui surumono (Now, What to Surround). Anutariainu: We Humans 6, 7 merger number: 4.

Sato, K., Tanaka, H., Aoki, J. and Horie, T. (2016). Okinawa Heripaddo: 'Dojin' Hatsugen Kidoutaiin ni 'Syucchou Gokurousama' (Okinawa helipad: 'Good Work' Response to Riot Policemen Making a Racist Slur 'Dojin'). *Mainichi Newspapers,* Oct 20.

Scott, S. (2013). Ainu Fight for Return of Plundered Ancestral Remains, *Japan Times,* Aug 12.

Segawa, T. (2015). *Ainugaku Nyumon (Introduction to Ainu Studies)* (Kodansha Gendai Shinsho, Tokyo) pp. 13–65.

Shigeru, K. (1994) *Our Land was a Forest: An Ainu Memoir* (Westview Press, Boulder).

Ueki, T. (2005). Kodama Sakuzaemon no Ainu Zugaikotsu Hakkutsu: Haikei to Gaiyou (Notes on the Excavations of Ainu Skulls by Kodama Sakuzaemon: Background and Outline), *Bulletin of Tomakomai Komazawa University* 14: 1–27.

Ueki, T. (2007). Kodama Sakuzaemon no Ainu Zugaikotsu Hakkutsu: Hakkutsu no Ronri to Rinri (Notes on the Excavations of Ainu Skulls by Kodama Sakuzaemon, Ethical and Socio-political Problems), *Bulletin of Tomakomai Komazawa University* 17: 1–36.

Ueki, T. (2017). *Gakumon no Bouryoku: Ainu bochi ha naze abakaretaka (Violence of Academics: Why was the Ainu Cemetery Excavated?)* (Syumpusha, Yokohama).

Ueki, T. (n.d.). *Syokumingaku no Kioku: Ainu Sabetsugaku to Gakumon no Sekinin (Memory of Colonialism: Ainu Discrimination and Academic Responsibility)* (Ryokuhu Syuppan, Tokyo) pp. 10–35.

Uemura, H. (2001). *Senjuminzoku no 'Kindaishi' ('Modern History' of Indigenous People)* (Heibonsya, Tokyo).

Winchester, M. (2014). Ainu Sengoshi no Bouryoku Hihanron: Ikotsu Mondai to Hashine Naohiko Saiban wo Tegakarini (A Critique of Violence in Ainu Postwar History: On the Ancestral Remains Issue and the Trial of Hashine Naohiko), *The Bulletin of the Research Institute for Japanese Studies, Kanda University of International Studies* 6: 59–89.

CHAPTER FIVE

Legacies of Eugenics, Race, Education and Colonization: Reflections on the San Carlos Apache Nation

*Darryl Macer**

Abstract

There have been debates in almost every corner of the globe over the definitions of culture, identity, and what constitutes ethics. These social constructs are a product of an individual's ontology, genes, environment and relationships. Colonization has been a major force to articulate bioethical value systems that were previously implicit in the relationships of people and the natural world. Ethical values and principles have developed in the context of epistemological systems and are central to how knowledge is gained and organized, how knowledge is used, and who has access to it. Through bioethics dialogue we will go beyond the legacy of colonization and eugenics, beyond European and Anglo-American norms, and stimulate education globally.

The fundamental question asked in this chapter is whether in the colonization of the Apache people, other Indigenous Americans, and Indigenous peoples generally, the people were killed because of competition over land, water and resources, or because of perceptions among the white colonizers of racial superiority, or because of ideological

---

*President, American University of Sovereign Nations, San Carlos, Arizona, USA; Director, Eubios Ethics Institute, Aotearoa/New Zealand, Japan and Thailand.

differences in religion and spiritual practices? The reservation system created a dependency upon external food and broke significant parts of the spirit of the Indigenous people. This made exploitation easier, and it has lasting ramifications in health, education and life choices. Economics and exploitation were significant causes, as was perceived white racial superiority.

A holistic definition of health is critical for human beings to flourish. The physical, social and emotional health of the Apache community is poor. Recommendations include (1) improving Native language programs so that all school students will be bilingual; (2) independent control for educational curriculum and development; (3) sovereignty over land and water and other environmental resources, including minerals; (4) improving self-esteem; (5) economic empowerment; (6) more research into the historical interface between eugenics, race and colonization of Indigenous American peoples; (7) bioethics education for empowering citizens to prepare for the policy challenges and implementation of genetic and other measures which may worsen the stigma, and discrimination, faced by Indigenous Americans given the new technological developments; (8) return to dependence upon traditional food and medicines as a primary source, supplemented by Western sources when they are better on a case-by-case manner; (9) lessened dependence on government control and outside sources; (10) independent governance free of corruption from business interests; and (11) instilling a love of life and pride in all people.

Colonization of Indigenous Peoples and the Apache

There are fundamental bioethical questions about the colonization of Indigenous peoples. This paper will focus on the Apache people, and in particular those who identify with San Carlos Apache reservation in Arizona, United States of America (USA). There are connections to other Indigenous Americans, as well as other Indigenous peoples [Newcomb 2008]. In particular there is a discussion of attitudes and experiences of vulnerable peoples during the 19th and 20th century. The Apache people are mostly located in Mexico and the USA. There are differences in the approaches to colonization of Apache seen in Mexico, where a majority of the dominant population are mestizo, persons with genes from both Spanish and Indigenous origin, compared to the approaches seen in the

United States where the majority of the dominant population from the 18th and early 20th century were of white European genetic origin.

Our psychological identity is affected by the heritage that we have, genetically, socially and environmentally. Genetic reductionism is the concept that many of our attributes are determined by our genes, including the percentage of blood that we share from tribal ancestors. In most tribes in the USA today, there are limits on the amount of genetic ancestry that we can trace to tribal ancestors, though this concept is under review in some tribes that are questioning genetic reductionism. Some tribes use the proportion of blood as a means to limit the number of persons in a tribe that are entitled to per capita payouts of tribal income from enterprises such as casinos.

We see important changes over time in the attitudes toward Indigenous peoples from the dominant populations also, and in the case of USA, Indigenous Americans were only legally recognized as US citizens in the 1920s. Some attitudes do not appear to have much logic beyond being justifications for exploitation. I will examine how the development of scientific methods for assessing racial purity, craniometry, and eugenics, affected the philosophy underpinning the relationships between the dominant white population and Indigenous Americans.

Early European Contacts with Apache People and Competion over Land Use

The term Apache is one adapted by the Spanish, but Apache people themselves use the term "nde", meaning the People. The Apachean or Southern Athapaskan language can be divided into seven tribal groups: Navajo, Western, Chiricahua, Mesalero, Jicarilla, Lipan, and Kiowa-Apache.

Because of the nature of their life involving seasonal migrations over large lands, the concepts of land, place and home, cover a broader range than those Peoples who follow a sedentary life and a fixed land use agricultural tillage pattern. In his 16th Century report, the Journey of Coronado, Castaneda described them as:

> people who lived like Arabs and who are called Querechos in that region...
> These people follow the cows, hunting them and tanning the skins to take

to the settlements in the winter to sell, since they go there to pass the winter, each company going to those which are nearest... That they were intelligent is evident from the fact that although they conversed by means of signs, they made themselves understood so well there is no need of an interpreter...They have better figures than the Pueblo Indians, are better warriors, and are more feared [Macer 2016].

There was an existing system of tribal movements and patterns of migration that allowed toleration of different tribes and people using the same space. We can see this also in other migratory peoples, including nomads in North Africa and the Middle East, today. That is also why they were described by the Spanish as "Arabs" of America in the 16th Century.

We can see evidence of the contacts between peoples in some written records. For example, at the end of the 15th Century the Mexican town of Monterrey was twice destroyed by the Apache tribes, before the third settlement was made to be resilient. In many areas however, the tribes were left to wander over their land. Gradually over two hundred further years the number declined, and more settlement occurred. There is considerable documentation, and they maintained a reputation as intelligent warriors. They also had skirmishes with many other tribes, as they had reportedly had prior to the Spanish. Their nomadic life would naturally have led to conflicts with tribes who had chosen to live in lands that these movements crossed.

In 1775 we see Lieutenant Colonel Hugo Oconor planned military operations against the Lipan and other Apaches in numerous locations. In 1776 we can read in a report to Teodoro de Croix, commandment general of the Provincias Internas, Apache depredations occurred in various places [Macer 2016]. There were also attacks by Comanches against the Lipanes. We can read of different approaches of Mexican generals, for example, "Colonel Juan de Ugarte is said to be inclined to pacific measures while Ugalde preferred the sword to do its work first" [Neighbours 1975, p. 35]. The skirmishes continued, and we read in a 1788 report to the King of Spain that bands of Lipanes were in the frontiers of Texas, Nueo Leon, among other places. The Spanish missions for the Lipanes on the upper Nueces were abandoned in 1771, presumably because of the reluctance of the locals to having a Spanish presence. Bubi had recommended that the Lipans be "exterminated", but they survived despite the attacks from the Spanish from the south and the Comanche, Wichitan and Caddoan from the north.

In 1820 the Lipanes were said to have suffered a defeat at the hands of Spanish troops on the Guadalupe River and a disastrous defeat at the hands of the Tawacanos on the Colorado Rover [Macer 2016]. After the Mexican revolution of 1821 there are less meticulous records than those the Spanish kept, but the Mescaleros were reported to have taken refuge near Chihuahua from the Comanche.

After that there was a peace treaty between the Lipanes and the Republic of Texas, that allowed the Lipanese to continue nomadic habits peacefully [Neighbours 1975, p. 39]. This general area was Southwestern Texas. In 1847 the former President of the Republic of Texas, David G. Burnet, reported that the Mescaleros, including between 1000 to 1500 warriors, were peacefully cultivating the soil on the Pecos. However, two years later, a Comanche raiding party entered the area, followed by troops which killed 30 of the Apache, who fled and then returned to a prior land into Texas. In 1850 United States Special Indian agent John E. Rollins reported that the Lipanes made corn on the Pecos while some were on the Rio Grande, with a population of 500 persons including 100 warriors. In 1851 another Special agent, John A. Rogers held a Council meeting with the Texas Indians.

Those Lipans living in Mexico faced attacks from Mexico, and the Mexican secretary of war recorded that the population of Lipan warriors had been reduced from 1000 in 1822 to 88 warriors in 1855. There were conflicts between both populations as evidence from the conflicts and decreased population.

Through the 1850s and 1860s there were claims of cattle being taken from cattle ranchers by Apache tribes, and also reports of shortage of foods, and territorial disputes. There were claims in the US Courts that the Apache had not been present in earlier lands, although their presence is reported in the Spanish documentation, which allowed the white settlers to take control of an increasing amount of land. As reservations were established, there continue to be reports of Apache raiders outside of the reservations into the 1880s.

We can see from these examples of the historical accounts that Apache people were attacked by colonizers of British and Spanish origin [Opler

1941]. Were these because of racism or simply resource competition? We can see from the earlier historical records of the Spanish that Apache peoples were widespread in the South West United States and Mexico, but they were increasingly being gathered into smaller land areas, ultimately into the reservations.

The conditions on the reservations were poor and are discussed below. It is therefore of no surprise that there would be attempts to gather medicinal plants, food and other items from the traditional lands. The bioethics of many Apache people today still have a more ecocentric focus compared to the dominant anthropocentric paradigm of the majority of people in the USA [Macer 2016]. For example, trees and animals are valued for their own sake, not just because of utility to humans. Water resources should be shared with animals and plants. The health of a society, and nation, includes the health of plants and animals — not just human beings. In fact, the public health of humans is blamed on their loss of the connection to nature, and practices such as harvesting of cactus that used to bring people out of their houses for exercise, a spiritual connection with nature, and healthier food than the dominant fast-food of today.

San Carlos Internment Camp

San Carlos, Arizona, was a destination for about fifteen tribes deliberately mixed together with the hope that if they did not die, they would inter-breed, and this might dilute the tribal identity [Macer 2016]. There were also a mix of persons who were working as scouts for the US. Army, along with those strongly opposed to colonization of land, and the confinement of a nomadic people. These tensions are still are present in the members of San Carlos Apache nation today [López Hernáez & Macer 2018].

The motivations of some of the officers and those who interned Apache and other tribes was not always war, although they belonged in the Ministry of War, and the "Apache Wars" are famous [Goddard 1919]. We can see some of the changing attitudes with a linkage to eugenic ideology in quotations. John G. Bourke [1891] wrote:

> The transformation effected was marvelous. Here were six thousand of the worst Indians in America sloughing off the old skin and taking on a new

life. Detachments of the scouts were retained in service to maintain order; and also, because money would in that way be distributed among the tribes.

However, Bourke [1891] was clearly against the establishment of the San Carlos reservation, writing:

> There is no brighter page in our Indian history than that which records the progress of the subjugated Apaches at Camp Apache and Camp Verde, nor is there a fouler blot than that which conceals the knavery which secured their removal to the junction of San Carlos and Gila.

There were a number of events, including group punishment, starvation and violent acts, that sowed seeds of discontent and mistrust, that is still found today, 140 years later.

The Indian Bureau chief, Agent Jon P. Clum, wrote to Washington that a steady stream of transfers of Apache bands from Camp Verde, Camp Apache, the Chiricahua Reservation, and Ojo Caliente, to concentrate them in San Carlos was successful and satisfactory. This was contrary to the opinion of the best judgment of the Army officers in command and opposed to the desire of the Indians [Macer 2016]. There were Grand Jury indictments of some Indian agents, such as Agent Tiffany for the confinement of 14 Apache men without charge for 14 months. Among citizens of Arizona territory there was apparently a wider understanding that fraud was common. The Indian agent is also reported to have sent Apache men to work mining coal in order to receive the rations, that were meant to be free. The profits were kept by some agents [Macer 2016]. There are positive reports of General Crook who returned in 1882 to try to repair the situation and trust that had eroded with the Army crimes at San Carlos.

When groups of Apaches left the reservations, they were called renegades, and their conditions were often poor. Some set up camps in the arid flat land. Others, especially in the hotter seasons of the year, would go to traditional land such as the forests on the 3,000 m high Dzil Nchaa Si' An [Mount Graham N.D.], or Oak Flat [Arizona Mining Reform Coalition 2009]. In doing so, they were pursuing the ethical principle of autonomy, self-rule, pursuit of traditional lands and medicinal plants and food sources which were generally healthier than the conditions in the reservations. Life was still tough, but having freedom is an ambition that we take as a basic human right today.

One of the most significant changes that was made through the reservation system was the loss of traditional foraging for foods and medicinal plants, instead being forced to rely on handouts of flour, oil and meat. This diet, as well as the dependency on handouts, was forced upon people who were punished if they left the reservation to forage and hunt for traditional foods. So instead of gathering foods, most modern Apache simply shop in a supermarket and eat fast food. Thus, the quality of life in traditional lands was better than in the reservations, and because the white settlers, farmers and mining companies wanted the better land they forced the original inhabitants off the land. This also inhibited the expression of religion and spirituality. Praying to the rising sun, and setting sun, and to all life forms, is a core belief of the Apache religion. The same is true of Shinto in Japan, and fortunately for the Japanese, their island location and distance, and degree of socio-economic development made it impossible for Europeans to colonize.

Health

The average life expectancy of Indigenous Americans on reservations in USA was 49 years of age, a 30-year gap with the average life expectancy in the United States in 2010. This is very disturbing ethically, and in the same way that many writers in the 19th century called the federal Indian policy a failed one, we could make the same observation today, in the 21st century. There have been reviews of the Indian Health Care service [Macer 2016], and why for various reasons it has failed to deliver adequate health care, both preventative and public health, and services to the communities that were promised high standards of health. The San Carlos Apache tribe established a private hospital and health system in order to attempt to improve delivery of health care to the community members.

A number of health conditions have multiple causes and simple models of determinism can lead to stigmatization. Diabetes and obesity are endemic on Native American reservations. The cause of this includes diet, genes, lack of exercise, and culture. Diet and exercise programs are needed to reduce morbidity and mortality, and social support systems to motivate persons to maintain these lifestyle changes.

The finding that there is an association between an allele of the gene for dopamine D2 receptor and alcoholism, illustrates the type of dilemma. There has been four decades of research which has shown that part of the vulnerability to becoming alcoholic after exposure to alcohol is inherited [Gordis *et al.* 1990; Blum *et al.* 1990]. Understanding how genes and environment interact to lead to alcoholism is a broader challenge. There are several genes involved, and many cultural factors. It is a major problem on the reservations, but it is determined by genes and culture, and the examples of families in the past. Because of research abuses there is suspicion of genetics and medical research, and close supervision nowadays through Institutional Review Boards, and sometimes Tribal Councils.

The American Eugenics Society changed its name to the Society for the Study of Social Biology in 1972. The social environment was thought by some, such as Muller, Huxley and Osborn to be one of the main directors of natural selection, and that eugenic goals could not be readily achieved in capitalist societies [Kevles 1985; Freeden 1979]. Capitalist society is dysgenic [Huxley 1936]. Osborn [1940] advocated a type of social welfare state to aid eugenics. The issues of eugenics and the ways it may be implemented in public policy are not just based on genetic ideas but consider the economic and social system.

The major use of eugenic selection occurred together with the move to a more scientific worldview. This is the result of both the development of scientific techniques, from sterilization operations, genetic screening to gene therapy in the immediate future; and from the associated cultural values. As our genetic knowledge greatly increases, we must note this tendency. We must be careful about the possible growth in genetic reductionism that could come from the detailed analysis of the human genome. This will be a challenge to existing human society, and it will need to be introduced slowly, in a way that is sensitive to any adverse social consequences.

Both Assimilation and the Reservation System Negatively Affected the Health of People

The *New York Times* in February 1880 wrote:

> The original owner of the soil, the man from whom we have taken the country, in order that we make of it the refuge of the world, where all men

should be free if not equal, is the only man in it who is not recognized as entitled to the rights of a human being [Hoxie 1984].

Here we can see an early understanding that the health of a person is the total of the physical, mental and spiritual aspects of a person.

Despite the government decisions to promote the reservation system we can find a number of voices against the system in the White Community. After massacres, in January and September 1879 for example, there was criticism of the reservation system. The *Alta California*, a San Francisco newspaper, in an editorial called the reservation policy a "murderous system", "it is starvation for the savage, it is oppression by the lawless white pioneer, it is death to our gallant officers and men". "One thing is certain, and that is that our whole Indian policy is a miserable one and a failure".

In late 1879 a Ponta Indian chief, Standing Bear, made a number of public appearances to large audiences in Chicago, Boston, New York, Philadelphia and Washington D.C. He condemned the reservation system and called for the extension of constitutional guarantees to Indians [Hoxie 1984]. In January 1880 the Commissioner of Indian Affairs (Ezra A. Hayt) resigned after accusations of corruption on an Arizona reservation. There were various Indian chiefs who spoke for ending the system, including John Ross, Black Hawk, Red Cloud, Geronimo and Sitting Bull. There were also various public groups and they were similar to some of the public campaigns in the 1840s and 1850s against slavery [Hoxie 1984]. The argument of assimilation would end reservations, and the compensation for loss of land would simply be to have full citizenship of the American society, a booming nation.

Emergence of American Anthropology

On 3 March 1879, Congress passed the annual civil appropriations bill including an authorization of USD 20,000 to establish a Bureau of Ethnology within the Smithsonian Institution. On 4 March 1879 the Anthropological Society of Washington listened to a paper by Frank Cushing, "Relic Hunting". The Bureau changed its name in 1893 to the Bureau of American Ethnography. There was then a lot of systematic research conducted on the American Indians, who were conveniently confined to reservations for the researchers. Unlike new migrants they were also living outside mainstream society, so they were politically safe as targets of research.

The founding Director of the Bureau of Ethnology is well known as an explorer in Arizona, Major John Wesley Powell. The predominant ideology was of social evolution, with books such as Henry Morgan's *Ancient Society* of 1877. The idea was that human history moved from simplicity to complexity, with a progression from savagery to barbarism and enlightenment. Many researchers and social engineers could also embrace the concept of assimilation of Indigenous Americans, and Indigenous persons globally, through the concept of social progress. The practice of anthropologists has raised numerous ethical concerns, and still there are conceptual gaps between anthropologists and concepts such as informed consent from Institutional Review Boards (IRBs). For those tribes that have formed their own IRBs, such as the Navajo nation, they have worked with tribal governments to limit the conduct of researchers upon tribal members.

From Citizens to Non-citizens and Non-persons to Citizens

This viewpoint of assimilation seems to be at odds with the eugenicists who were also emerging at this time; they argued that the brain of the white person was superior to the brain of the black or "American Indian". Eugenics is discussed in further detail later. In the case of colonization, a more important question is that of citizenship. The original inhabitants of a land can be described as the citizens of that land. This is not decreed by a written piece of paper, but by the entwinement of the people with the land throughout their history.

Ironically the colonizers to the United States decided that the original inhabitants were not citizens. In fact, in the 19th century and early 20th century the term "Native" was applied to US. citizens, of all races, in comparison to those people who wanted to be immigrants. Thus, we can see how the white government preferred to call the Native Americans or Indigenous Americans "American Indians", because it was convenient to call themselves, the new landowners, the natives. The immigrants who came in the 20th century (and since) had to pass a quality assurance program, that the earlier people did not. There were various eugenic and racial measures applied in immigration policy [Macer 1990].

By 1879 there was a resurgence amongst white Americans that Indigenous Americans should be considered citizens under the law. In the early 20th century, racial superiority and social progress ideals arguments were still being used as a reason not to grant Indigenous Americans full US. Citizenship, i.e. by claiming that they were a "backward race", and arguing that they would abuse their full US citizenship, or be taken advantage of by the unscrupulous [Hoxie 1984].

Finally, in 1924 all Indigenous Americans were granted full citizenship, after these types of debates, in the 1924 Indian Citizenship Act 43 [Macer 2016]. This is obviously counter to bioethics and human rights, but consistent with practices such as bounty hunting to reward hunters who could bring a redskin scalp to the government office after killing Native Americans in the 19th century.[1]

Education for All to Become "Civilized" was Counter to Health Promotion

Education through community learning has been a norm for all human beings through evolution and is critical to our health. This includes education in hunting, fishing, raising children, music, prayer, and selecting food, for example. Particular cultural patterns and traditions reinforced this, and culture is closely tied to education. There were education systems in a number of Indigenous communities prior to European colonization, and we can see the famous remains of academies in Incan, Aztec and Mayan culture. Advanced astronomical knowledge was critical for the marking of solar and lunar solstices that were critical to agriculture and survival.

Today, as in the past, national governments set curriculum that are often tightly linked to the construction of a society where individuals think in a particular manner. The Ministry of Education, or in the United States case, the Department of Education, is often conservative in their values.

---

[1] For example, in the *Daily Republican* newspaper, Winona, 24 September 1863, p.1, the bounty was reported to be increased to $200 per person killed.

Organized education is a type of social engineering, and nowadays we still hear elders complaining that the young generation are just not being taught properly in school.

The boarding school experience for Indian children began in 1860 when the Bureau of Indian Affairs established the first Indian Boarding School on the Yakima Indian Reservation in the state of Washington. These schools were part of a plan devised by eastern reformers Herbert Welsh and Henry Pancoast, who also helped establish organizations such as the Board of Indian Commissioners, the Boston Indian Citizenship Association and the Women's National Indian Association. The goal of these reformers was to use education as a tool to "assimilate" Indian tribes into the mainstream of the "American way of life," a Protestant ideology of the mid-19th century. Indian people would be taught the importance of private property, material wealth and monogamous nuclear families. The reformers assumed that it was necessary to "civilize" Indian people, make them accept white men's beliefs and value systems. By the 1880s there were over sixty boarding schools, and many children were forcibly taken away from their parents and their home communities, in efforts to make them confirm to a white model of a person. This made it more difficult for the youth to be educated in traditional patterns of health and diet, and also led to estrangement from the environment.

Although education was being offered to all Native American persons, through a Boarding School system, the education was to provide persons as servants of the upper class of the colonizers. Trades schools to produce, maids and waiters, were for the more feeble-minded races. In most cases children of Native American families were forcefully taken from their families to boarding schools. Many children died at the schools, and many lost their identity through these assimilation attempts, which included forbidding children to use their mother tongue. The graveyard was one possible outcome, and not so many graduated.

Eugenic ideology would state that a person of a certain genotype can only work in a certain occupation and profession, i.e. genes determine your future occupation. All forms of racial discrimination are unethical [UNESCO 1950; Macer 1990]. By the end of the 20th century there had been recognition of the greater importance of non-genetic factors in determining intelligence, criminality and social desirability. It has been found that the trend in the USA for family size to be decreasing can be correlated with an increased level of educational attainment. The fewer the number of siblings, the higher chance of continuing education. On reservations however, there were large family sizes, many teenage pregnancies and high levels of violence. The development of the American Indigenous Research Association in 2013 is some indication that there is progress in the movement to elevate Native Americans from being merely the subjects of research to being researchers themselves [Swisher 2004]. There have also been questions on the ownership of knowledge in the research [Archuleta 1990]. At least we can see success in a growing number of Native Americans with doctorates and Departments of Native American studies at some universities.

No doubt there have been individual successes, and also some progress in the Apache language programs so that the language stays alive in the community, but high mobility and dropout rates continue to plague the youth today. An extensive review of the San Carlos Apache tribal educational system of the 1950s and 1960s [Parmee 1968], reveals the same problems that exist today. The hopes and aspirations of the tribal members, educationalists and others expressed then are similar to those of today. We need to find different solutions and learn from the obvious lessons.

Given the theme of this paper, what is interesting is that there is a shift in ideology of the Indian Boarding Schools, which had intended Native American students to only work in service industries because of either a perceived low intelligence, or the desire to maintain the Native American tribes as suppressed communities of people who would not challenge the ruling elite. The 1960s was considered a revolutionary time in development of human rights, and saw the termination of the policy in the 1950s to assimilate tribes into white mainstream culture. The 1960s saw the rise of self-determination in Native American tribes [Macer 2016]. There is evidence for both ways of thinking. Parnee [1968] wrote:

> any community — when it is manipulated by outside sources and its people are neither trained nor given an increasing share of the responsibility for their own affairs, when they are deprived of their traditional heritage while pressured to accept change, and when their social, political, and economic institutions are disrupted without provision for immediate or adequate replacement.

There were significant attempts to demand uniformity in the Indian Boarding Schools. The presence of uniforms and strict discipline was common to many schools at the same time. The most serious impact however was the ban on speaking languages other than English, so that the children would be discouraged/punished if they spoke Apache. This led to a loss of identity and conformity. This cultural and identity confusion, through multiple generations, has significant psychological and developmental problems. The eugenics rhetoric was dominant and remember that these children would only become US citizens in 1923.

Many scholars have examined the curriculum, teaching methods and goals of education for Native American students [Ryan & Brandt 1932; Dumbleton & Rice 1973]. There continues to be academic discussions of how to improve tribal education because the results continue to be disappointing.[2] Anderson [1981] argued that the expectations of the students, the teachers and the parents were all different, and each group did not know what the other's expectations were. Some of the concerns, such as a system that does not encourage creativity, are more universal. Huxley [1932] wrote that particular goals for each tribe should be used for deciding the type of education of the people, 13 years before he would become the first Director-General of UNESCO. He encouraged art, pottery, basket weaving, as well as other culturally orientated educational goals that would reinforce the community values.

Eugenics

The word eugenics, "good genes", may be separate from the very common view that the mating of people of "good views" is desirable, to give us more offspring of that view, but we will see that eugenic proponents have often retained this idea [Macer 1990]. This interplay between the concept of possessing good genes or possessing good views, is very relevant to the various policies used in the interactions of European colonization into lands occupied by Indigenous people. We see for example, cases

---

[2] For example, see a special issue of *Peabody Journal of Education* 61(1), 1983. Of course, there are many contemporary articles, but as illustrated these discussions have continued for a century — without a solution for education for many students who still cannot seem to excel.

where extermination campaigns were made against some Native Americans, and Tasmanian Aborigines, which were followed up with assimilation campaigns to take the children of Indigenous persons into either new families (forced adoption into white families), or into long term Boarding Schools.

The 1870s were an interesting time for this discussion and also when the San Carlos Apache reservation was established; in the same decade the eugenic paradigm saw a resurgence. The concept of social progress and the American Ethnography Society were established. A major motivation underneath many eugenicists was also the idea of human progress, that we must be progressing genetically as well as in our knowledge. This was boosted by the theory of evolution, the survival of the fittest was equated with the survival of the "best". The best were the best people to cope with modern life. Galton was a cousin of Charles Darwin. Social Darwinists tended to equate a person's genetic fitness with his social position. Social Darwinist ideology provided a good climate for eugenic thought, and many qualities such as intelligence, temperament and behavior were believed to be inherited [Ludmerer 1978; Hansen & Kind 2013].

Dowbiggin [1997] documents the involvement of both American and Canadian psychiatrists in the eugenics movement of the early 20th century. Psychiatrists at the end of the nineteenth century felt professionally vulnerable, because they were under intense pressure from state and provincial governments and from other physicians to reform their specialty. Eugenic ideas, which dominated public health policy making, seemed the best vehicle for catching up with the progress of science. Prominent psychiatrist-eugenicists included G. Alder Blumer, Charles Kirk Clarke, Thomas Salmon, Clare Hincks, and William Partlow [Allen 1989]. Psychiatrists played roles in the tough debates about immigration policy.

One of the most significant attitudinal changes made in the process of colonization was the shift from being self-sufficient in food provision, to being dependents because the Apache people were confined to reservations. Given the belief in dependency as a motive for eugenic sterilization, it is relevant to consider the emergence of dependency as a trait in Apaches who were admired in the 16th century by the Spanish and were clearly self-sufficient. By the late 1800s there were queues of people for rations. Rations of flour, sugar, coffee, and meat were given out every ten days.

Since there were many people it took all day to get the food. Later instead of beef the people were provided cattle to butcher themselves, with ten to fifteen head of cattle per band. If the Apache were allowed to leave the reservation to forage and hunt as they had, they would not need to be relying on the food — given that this food, supplemented with lard would lead to an unhealthy diet and eventual obesity. This is illustrative of the issue for many tribes forced to stay on reservation lands.

Sterilization Programs

The United Nations World Population plan of action declares that, "all couples and individuals have a basic right to decide freely and responsibly the number and spacing of their children". Eugenic policy had a very serious human rights abuse with the sterilization programs of Native Americans through the Indian Health Service in the 1970s [White 1978]. In the 1970s for example, the forced sterilization was administered under the false pretext of laws to eliminate poverty [Torpy 2000].

Some of the conditions thought to be heritable were "nomadism", "shiftlessness", and "thalassophilia" (love of the sea) [Haller 1963]. The American Eugenics program was tied to the European programs. You can see that nomadic tendencies of Native Americans would place their traditional life choice as one which was an indication for sterilization [Bruinius 2013]. In 1935 the American Eugenics Society produced a major work called "Tomorrow's Children" [Huntington 1935]. It expanded the number to five million adults and six million children who were "subnormal in education", and another twenty million who failed to finish grammar school [Mehler 1987]. It recommended that these people should not be allowed to breed, though not all should be sterilized. There were, and continue to be, very low graduation rates for Native Americans. Under education criteria they would therefore be in the groups subject to sterilization. During the 1960s and 1970s

the Indian Health Service deceptively sterilized a large percentage of Native women who were between the ages of fifteen and forty-four. Those doctors failed to provide women with necessary information regarding sterilization; used coercion to get signatures on the consent forms; gave improper consent forms; and lacked the appropriate waiting period (at least seventy-two hours) between the signing of a consent form and the surgical procedure [Indian Country Media Network 2015].

England [1994] discusses the events and studies made of the 1972–1976 sterilization of Native Americans, which included tribal members across Arizona, and the whole United States. The abuse of power by the coercive sterilization programs was called genocide by many commentators and implies that the responsible persons working in the Indian Health Service still had a belief that the Native American women were inferior in intellect and had to be sterilized. This patronizing attitude is still seen today in my observations of Indian Health clinics and the population. Racist attitudes may also have been a cause of these sterilizations in both Canada and the USA [White 1978].

Although sterilization of Native Americans had still continued after World War II, the general eugenic sterilization programs went into decline in the 1940s, largely because of the eugenic excesses of the Nazis [Petchesky 1979]. However, sterilization operations were still performed after the war, particularly in Georgia and North Carolina in the USA [Reilly 1987]. In Alberta, Canada, the Sexual Sterilization Act (1928–1973) concentrated on Indians and Ukrainians, and 25% of the women sterilized were First Nations people although they only make up 3% of the provinces' population [White 1978].

In conclusion we can see this serious bioethical abuse in the decade that many Americans claim to be the first decade of modern bioethics — that between 1970–1975, physicians sterilized about one quarter of all Native American women of child-bearing age through the Indian Health Service [Kluchin 2009]. Can we really still see the USA as the birthplace of bioethics?

Conclusions

A holistic definition of health is critical for human beings to flourish. The physical, social and emotional health of the Apache community is poor. Colonization has led to loss of land, loss of control of water, loss of buffalo and traditional food sources, changed eating and living habits, as well as a transformed and divided religious identity. The question of ongoing federal financial support for Native Americans is one which should be explored with regard to the economic arguments so often applied to eugenics [Thompson 1979]. Given the amount of land, water and mineral resources taken from Native American land, there could be no ethical justification made to limit financial assistance to Native Americans, who were promised free health care and free education in return for acceptance of the reservation system and giving up land to white settlers. We need decolonization of education, health, land and ideas. The American Indian Religious Freedom Act (AIRFA) was signed into law on August 11, 1978 in the United States. The Act was created in an attempt to correct the wrongful treatment of Native Americans and their religious beliefs. Ironically, even today, many tribal members are still under pressure of forced assimilation and conversion to the Christian way of thinking, even by Christian Native Americans. The issue is still divisive, and continues to separate families today. The stated goal is, "The Act is intended to guarantee to native peoples — American Indians, Native Alaskans and Native Hawaiians — the right to believe, to express, and to practice their native traditional religions" [Native American Rights Fund 1979]. During this time there were also many different religious groups encroaching on reservations trying to convert the people to different ways of worship [Bighorse 2016; Harjo 2004].

The American Indian Religious Freedom Act came with stipulations. Even though it is based on the First Amendment of the United States constitution, as the freedom to exercise one's religion, the use of sacred sites is only to the extent that is feasible and is not for proposals contradictory with any United States government operation. Native American Indian tribes all over the United States are fighting for their inherent rights to exercise their cultural beliefs on their own traditional homelands and the United States government is disrespecting tribes by requesting permits for sacred objects and access to sacred sites, some of these based on mining, such as Oak Flat sacred site for the Apache community, or the oil pipe line in South Dakota. In early 2018 the Oak Flat spiritual site was vandalized, which has been classed as a "hate crime" by the FBI, and thus the bioethical issue of self-determination, as also articulated in article 1.1. of the United Nations Convention on Economic, Social and Cultural Rights, and of the United Nations Convention on Civil and Political Rights, remains central to the Apache people today.

Acknowledgments

I appreciate all those who have taught me lessons of life. I owe thanks to my colleagues who are professors and students of the American University of Sovereign Nations (AUSN), from all around the world, as well as to the members of Indigenous American communities who have been patient with me and have shared their wisdom and experiences with me.

References

Allen, G.E. (1989). Eugenics and American Social History, 1880–1950, *Genome* 31: pp. 885–889.

Anderson, N. (1981). The Teacher-Student Relationship in Apache Education, *Presentation at the Inter-tribal Parent Education Meeting, 30–31 January 1981.*

Archuleta, J. (1990). Discursive Analysis of Tiwa (Peubo) Text and the Ethical Question of "Ownership" of Research, *Redink* Fall 1990: p.7.

Arizona Mining Reform Coalition (2009). Oak Flat Land Exchange. Available from: < http://azminingreform.org/oak-flat-land-exchange/>.

Bighorse, L. (2016). *The Holy Fight Defending Oak Flat.* Masters in Bioethics and Global Public Health thesis, American University of Sovereign Nations.

Blum, K., Noble, E. P., Sheridan, P. J., Montgomery, A., Richie, T., Jagadeeswaran, P., Nogami, H., Brigggs, A. H. and Cohen, J. B. (1990). Allelic Association of Human Dopamine D2 Receptor Gene in Alcoholism, *Journal of the American Medical Association* 263: pp. 2055–2060.

Bourke, J. G. (1891). *On the Border with Crook* (Scribner's, New York).

Bruinius, H. (2013). *Better for All the World. The Secret History of Forced Sterilization and America's Quest for Racial Purity* (Alfred A. Knoff. New York).

Downbiggin, I. R. (1997). *Keeping America Sane. Psychiatry and Eugenics in the United States and Canada, 1880–1940* (Cornell University Press, Ithaca).

Dumbleton, D. D. and Rice, M. J. (1973). *Education for American Indians: A Book of Readings.* Anthropology Curriculum Project, University of Georgia.

England, C. R. (1994). A Look at the Indian Health Service Policy of Sterilization, 1972–1976, *Redink* 3 (Spring 1994): pp. 17–21.

Freeden, M. (1979). Eugenics and Progressive Thought: A Study in Ideological Affinity, *History Journal* 22: pp. 645–671.

Goddard, P. E. (1919). Myths and Tales from the San Carlos Apache, *American Museum of Natural History*, Vol XXIV, Part 1: p. 90.

Gordis, E., Tabakoff, B., Goldman, D., and Berg, K. (1990). Finding the Gene(s) for Alcoholism, *Journal of the American Medical Association* 263(15): pp. 2094–2095.

Haller, M. H. (1963). *Eugenics: Hereditarian Attitudes in American Thought* (New Brunswick, London).

Hansen, R. and Kind, D. (2013). *Eugenics, Race, and the Population Scare in Twentieth Century North America* (Cambridge University Press, Cambridge).

Harjo, S. (2004). American Indian Religious Freedom Act after Twenty-Five Years, *Wicazo Sa Review:* 19(2), pp. 129–136.

Hoxie, F. E. (1984). *A Final Promise: The Campaign to Assimilate the Indians, 1880–1920.* (University of Nebraska Press, Lincoln).

Huntington, E. (1935). *Tomorrow's Children: The Goal of Eugenics* (Wiley, New York).

Huxley, J. (1932). The Education of Primitive Peoples, *Progressive Education* IX (2): pp. 122–130.

Huxley, J. S. (1936). Eugenics and Society, *Eugenics Review* 28(1): pp. 11–31.

Indian Country Media Network (2015). Available from: <http://indiancountrytoday medianetwork.com/2015/10/03/forced-sterilizations-native-women-and-republican-attempts-shut-down-planned-parenthood>.

Kevles, D. J. (1985). *In the Name of Eugenics* (Knopf, New York).

Kluchin, R. M. (2009). *Fit to Be Tied: Sterilization and Reproductive Rights in America, 1950–1980* (Rutgers University Press, New Brunswick).

López Hernáez, L. and Macer, D. R. J. (2018). *Education, Happiness, Mindfulness and Colonization: Reflections from Time with San Carlos Apache Nation* (Eubios Ethics Institute, Christchurch, N.Z.).

Ludmerer, K. M. (1978). History of Eugenics. In Reich, W. T. (ed.) *Encyclopedia of Bioethics,* Vol. I (Collier Macmillan, New York) pp. 457–462.

Macer, D. R. J. (1990). *Shaping Genes* (Eubios Ethics Institute, Christchurch, N.Z).

Macer, D. R. J. (2016). *Legacies of Eugenics, Race and Colonization: Reflections on San Carlos Apache Nation*, Masters in Public Health thesis, American University of Sovereign Nations.

Mehler, B. (1987). Eliminating the Inferior, *Science for the People* 19(6): pp. 14–18.

Mount Graham Coalition (N.D.) Mount Graham Coalition Home. Available from: <http://mountgraham.org/>.

Native American Rights Fund (Winter 1979). We Also Have a Religion, *The American Indian Religious Freedom Act* and the Religious Freedom Project of the Native American Rights Fund. Announcements, pp.1–19.

Neighbours, K. F. (1975). *An Ethnohistorical Report, together with supporting exhibits relating to the cause of action in the Lipan Apache tribe, the Mescalero Apache tribe et al. v. United States, Docket No, 22-C, before the Indian Claims Commission.* In Apache Indians Volume xxx.

Newcomb, S. (2008). *Pagans in the Promised Land: Decoding the Doctrine of Christian Discovery* (Fulcrum Publishing, Golden, CO).

Opler, M. E. (1941). *An Apache Life-way: The Economic, Social, and Religious Institutions of the Apache* (University of Chicago Press, Illinois).

Osborn, M. E. (1940). Preface to Eugenics. In C. J. Bajema (ed.) *Eugenics, Then and Now* (Dowden, Hutchinson and Ross, Strondsberg) pp. 270–282.

Parmee, E. A. (1968). *Modern Apache Indian Community and Government Education Programs* (University of Arizona Press, Tuscon).

Petchesky, R. P. (1979). Reproduction, Ethics, and Public Policy: The Federal Sterilisation Regulations, *Hastings Center Report* 9(5): pp. 29–41.

Reilly, P. R. (1987). Involuntary Sterilisation in the United States: A Surgical Solution, *Quarterly Review of Biology* 62: pp. 153–170.

Ryan, W. C. and Brandt, R. K. (1932). Indian Education Today, *Progressive Education* IX (2): pp. 93–94.

Swisher, K. G. (2004). Pursuing Their Potential: TCUs Turn from Being Researchers to Being Researchers, *Tribal College Journal* 16, pp. 8–9.

Thompson, M. S. (1979). Prenatal Diagnosis and Public Policy. In A. Milunsky (ed.) *Genetic Disorders and the Fetus. Diagnosis, Prevention, and Treatment* (Plenum Press, New York) pp. 637–665.

Torpy, S. J. (2000). Native American Women and Coerced Sterilization: On the Trail of Tears in the 1970s, *American Indian Culture and Research Journal* 24(2): p. 5.

UNESCO. (1950). Statement on Race. In *The Race Concept: Results of an Inquiry* (Greenwood Press 1970), pp 98–103.

White, P. (1978). Forced Sterilization Amongst American Indian Women, *J. Indigenous Studies* 1(2): pp. 91–96.

CHAPTER SIX

Indigenous Collectivity and Bioethics in the Postcolonial Independent Nation of Tuvalu

*Eselealofa Apinelu**

Abstract

This chapter seeks to illustrate how the disconnection in individual rights law and collective interdependency is a source for bioethical concerns in Tuvalu. Collective interdependency is a Tuvaluan belief that the successful development of a society depends on respect for the duties, responsibilities and obligations (DRO) of individuals who each have a role to play and which are culturally shaped. While not overly deterministic, collective interdependency is an Indigenous lens through which Tuvaluans perceive, understand and experience rights.

Tuvalu is a group of islands in the Pacific Ocean where there is an ongoing fundamental tension in Indigenous understandings of collective interdependency and the law's more individual human rights approaches. Arguably, Tuvaluan beliefs in collectivity continue to be assessed through a colonial Western analytical lens, one that continuously doubted the overall merits of collective interdependency. This chapter uses the case of *Nukufetau v Lotoaa to* show how embedded collectivity is in Indigenous understanding and how the failure to properly incorporate Indigenous understandings in law, for a society where the

*Attorney-General and Indigenous Rights Strategist, Government of Tuvalu, Funfuti, Tuvalu.

importance of the collective remains a strong positive cultural value, is a cause for health challenges in Tuvalu. The paper introduces *Fenua* as an Indigenous lens for understanding collectivity.

This chapter argues that *Fenua* is core to health and wellbeing in Tuvalu. Where the law weighs heavily in support of individual rights, it impacts negatively on *Fenua* or the collective and individual health and wellbeing. This chapter concludes that, while respecting the autonomy and dignity of an individual, in societies where the importance of the collective remains a strong positive cultural value, there is a pressing need to reconsider prioritizing Indigenous collectivity in law.

Introduction — The Land

You must be in Tuvalu to believe that it is indeed a place on earth. You will not see it from the sky by plane until the final announcement of the descent from the cockpit of an ATR 72-600. It is that group of nine islands one barely notices on descent after a two-hour flight north east of Fiji on an ATR 72. It is not Tobaloo, Suva nor Vanuatu. It is pronounced Too — vaa — loo. Oh, and the landing will only take you to one of the islands in the group, Funafuti, the capital of Tuvalu.

For those who prefer geodetic datum for the whole group of islands, the nine coral islands of Tuvalu lie between 5 and 11 degrees south of the equator and just west of the International Date Line. The total land area is 26 square kilometers. One could easily walk the length and breadth of any island in less than a day. The only thing stopping such a doable task is that the tiny islands are spread over an ocean of 1.3 million square kilometers [Nations Encyclopedia 2018]. The highest point on the islands is about three meters above sea level [Government of Tuvalu 2012, p.7]. For those who have maps and have difficulty locating Tuvalu, try its colonial name, Ellice islands of the Gilbert and Ellice Islands Colony (GEIC).

The People

Tuvaluan people look no different from other human beings on the globe. In terms of physical appearance, you can find a Tuvaluan that resembles any stereotypical race on earth. Yes, *palagis* or Europeans, Africans, Asians, Islanders, Indians, and so forth. While looks may be deceiving, there is no deception in Tuvaluan peoples' connection to their *Fenua* or island. Tuvalu comprises eight self-governing islands, Nanumea, Nanumaga, Niutao, Nui, Nukufetau, Funafuti, Vaitupu and Nukulaelae. Niulakita, the most southerly of the group of islands is administered as part of the northern island of Niutao.

Indigenous Beliefs

Tales of how the islands of Tuvalu were originally settled differ between the eight islands of Tuvalu. The most northerly island, Nanumea, is thought to have originated from a Tongan warrior named Tefolaha [Isako 1983]. The people of Niutao in the north believe they originated from a spirit named Kulu [Nia 1983].

Nanumaga people believe they are descendants of Lapi, a Tongan who was brought to the land by a Fijian Sea Serpent [Lafai 1983]. The people of Nui believe they originated from Samoan settlers who came in a canoe named *Vakatiumalie* [Pape 1983]. Vaitupu and Funafuti are also believed to have originated from Samoan settlers [Laupepa 1983; Tinilau 1983].

Nukufetau is the only island with no celebrated tale of being settled or originating from foreigners. Could they be the true Indigenous Tuvaluans? Their only celebrated tale was that of the Island Chief Lagitupu who adopted and accorded a Tongan Stowaway, Laka, with chiefly status. Nukulaelae and Niulakita are believed to have been settled by people from the other islands of Tuvalu, Vaitupu and Niutao respectively [Faaniu 1983]. Every island speaks a different dialect with Nui speaking a totally different language which is more Gilbertese.

Prior to colonization, each island had sovereignty over all the affairs of its island including people. This same sense of island independence reflects strongly the varying cultures of each island that continue to form the basis of questioning colonial influenced laws in Tuvalu.

History, Custom and Colonization

Tuvalu was a by-product of British colonialism. It was colonized by Britain for 86 years from 1892 to 1978, [Faaniu 1983, pp. 185–186]. During colonial times Tuvalu was known as the Ellice islands. Prior to colonization the group consisted of eight independent islands with their own governance systems based on their Indigenous ways of perceiving, understanding and experiencing the world. The islands operated independently of each other save for the trade friendship forged during times of famines, traditional challenges and chance meeting up during journeys in the vast Pacific Ocean.

Following independence from Britain, the group became known as Tuvalu. It is the conjoining of all islands under the umbrella name of Tuvalu, eight united islands, for better or for worse, a colonial induced legacy formally accepted by the people of Tuvalu in 1978 when they became an independent nation. While Tuvalu may have become independent from British rule in 1978, its laws and formal system of governance continue to follow colonial influenced laws. Its culture, while anchored in island customs and traditions of collectivity, is strongly influenced by colonial remnants and globalization. Tuvalu may be what Loomba [2015, p.12] asserted as a country that is both postcolonial in being "formally independent" as well as inherently neo-colonial due to "economic and cultural" reliance.

Colonial Legacy

Colonialism has had a mixed effect on Tuvalu. One of the best impacts was uniting the independent eight islands of Tuvalu. One of the more problematic aspect was the disconnection or lack of synch in custom and law. Laws are written in a foreign language with foreign concepts. This disconnection is a fundamental tension and a bioethical challenge for Tuvalu.

Tuvalu Bioethics: A Case Study

Health is a "State of complete physical, mental and social wellbeing, and not merely the absence of diseases or infirmity" World Health Organization (WHO).

The case of *Nukufetau v Lotoala* is important to discuss for it shows how Indigenous understanding of collectivity is embedded and embodied in *Fenua*. *Fenua* is core to health and wellbeing in Tuvalu. Distorting this embeddedness and embodiment by operation of law is a nationwide bioethical issue, specifically for when decisions are made in favor of the individual, they impact negatively on both the collective and individual health and wellbeing.

Health, while commonly understood and discussed in terms of presence or absence of disease, also includes mental and social wellbeing. What is less discussed and understood, and is the focus of this chapter, is the mental and social health challenges caused by the disconnection in law and custom for societies where the collective remains a positive cultural value. Still less understood is that Indigenous understandings of collectivity extend beyond the Western understandings of separation of powers doctrine. Accordingly, in Tuvalu, custom and law is a health issue that bioethics must embrace.

A Glimpse of *Fenua* — Indigenous Lens

The tension between the operation of the law in Tuvalu and Indigenous understandings of collectivity is the focus of discussion in the chapter. The analysis is not a legal one, for to do so is to continue assessing the challenges from a colonial lens. Rather the chapter provides a social and cultural analysis of the health challenges from an Indigenous platform of collectivity I call *Fenua*, reflected through Nukufetau's journey in the *Nukufetau v Metia* case.

This paper does not in any way attempt to decide on the merits of the court decisions nor of the sentiments of the parties involved. Rather, it provides the reader with a more culturally authentic journey into the Indigenous world of *Fenua* in Tuvalu.

The Law and *Fenua* — *Nukufetau v Metia* [2012] TVHC 8

In 2012 the *Fenua* Nukufetau unsuccessfully petitioned the Court to remove Lotoala as a member of parliament for its constituency. The events leading up to this court challenge were believed by some as cause for civil unrest. To others, concern for civil unrest is almost a slap on the face of Indigenous collective interdependency. The court decided, unsurprisingly, for the law. Unfortunately, the polarizing beliefs in the same law between custom and individual rights would only exacerbate the tension between those for *Fenua* and those against *Fenua*.

Will there ever be real peace for the communities? Social problems and moral failures are closely linked to health problems [Behren 2013, p. 33]. One is always hopeful for positive results but for now why did it get to this? Why would the *Fenua* Nukufetau have to go to Court? Is it not the prerogative of *Fenua* to decide on matters concerning its island and subjects? This is but one of the disconnections in understanding between custom and law that is affecting people's livelihoods on the islands.

Factual Background — Aspiration for Leadership

The *Fenua* Nukufetau had high hopes for its representatives to hold top positions in parliament. It is a common aspiration of the islands in Tuvalu for their representatives to hold the considered "esteem positions" of Prime Ministers, Speakers or Governor Generals (*Sione v Teo* 2004). It is an understandable aspiration too given the background of the cohort of candidates Nukufetau had. All its candidates have had great experience in governance and had, at different stages of their careers, all served as Permanent Secretaries in the Tuvalu Public Service.

Enele Sopoaga had just completed tenure as Ambassador for Tuvalu in various overseas missions. Lotoala Metia was the Auditor General for the country before switching to politics. He was the Minister of Finance at the material time. Another prominent son of Nukufetau, Mr Elisala Piita, was one of the first two Tuvaluans to gain a Bachelor's degree during colonial rule.

In terms of Western qualifications, they all held prestigious qualifications. In terms of *Fenua*, they were equally respected elders of *Fenua*. It was a narrow call but Enele and Lotoala were the successful candidates for the 2010 general elections. A success soon seen as a blessing in disguise for both candidates who would have to endure so much of the complications of the disconnection between *Fenua* and law that were to follow.

Nukufetau — Lack of Esteem Accolade

Unfortunately, for Nukufetau, none of her representatives held any of the prestigious or esteemed positions following the 2010 general elections.

Lotoala and Enele joined different Prime Ministerial groupings. Enele's group formed the government but Enele was not the Prime Minister. After less than three months in office, the government was ousted by the successful passage of a motion of no confidence, moved by none other than Nukufetau's representative Lotoala. A new government with the support of Lotoala was formed but Lotoala was also not the Prime Minister.

The Vaitupu Connection

In 2011, Taukelina, one of the parliament representatives of the island of Vaitupu unsuccessfully brought an election petition against the other member of Vaitupu, Apisai Ielemia. One of the grounds for such petition was that Apisai had inappropriately used his position as PM to his benefit in the 2010 General elections. While the Court held that there was lack of evidence in support of such a petition, Nukufetau still considered Apisai's actions as morally, culturally and legally unacceptable. Nukufetau did not want any of its members to be associated with Apisai. Therefore, when Lotoala remained in the Prime Ministerial grouping that included Apisai, his action was considered one of defiance of Nukufetau's Indigenous understanding of DRO.

The Vaitupu case of Taukelina Finikaso and Apisai Ielemia (*Finikaso v Ielemia* [2011] TVHC 3) ran parallel with the case of Nukufetau v Lotoala (*Nukufetau v Metia* [2012] TVHC 8). While these two cases may be independent of each other, the outcomes support the same aspirations. A win to Nukufetau or Taukelina would increase the chances of getting the results that Nukufetau wanted. A win to Lotoala or Apisai would further dampen the aspirations of Nukufetau for national leadership. Nukufetau was represented at both spectrums of the political divide with Lotoala Metia and supporters on one end and Enele S Sopoaga and Nukufetau on the other.

Blurring the Issues

Although the Vaitupu case is separate from the Nukufetau case, I doubt most people, especially from Nukufetau cared for the difference. The thinking was that if Taukelina's case was successful, Nukufetau's aspiration for Prime Ministership could materialise. Mixed into all these cases was a defamation suit by Lotoala (*Metia v Finikaso* [2012] TVHC 2) against the initial supporters of the Maatia groupings which included the other representative of Nukufetau, Enele. While the defamation case may have had silent impact on the other constituencies, it exacerbated matters on Nukufetau. Enele had become the hope of Nukufetau for any chance at holding the esteemed position of Prime Minister. Lotoala, in continuing to be in a grouping Nukufetau did not approve of, was considered to be acting in defiance of *Fenua* and his challenge on Enele thus perceived as a challenge on *Fenua*.

While in law the *Metia v Finikaso* case and the *Finikaso v Ielemia* cases had nothing to do with Nukufetau, in custom, as far as Indigenous understanding is concerned, they are crucially related to Nukufetau's concerns for the role of Prime Minister. Nukufetau believed it was not right for her representative Lotoala to be in Apisai's grouping for Apisai had a case pending in court. Similarly, the successful defamation case by Lotoala in the *Metia v Finikaso* case, while only challenging Enele and others in their individual capacities and not as representatives of *Fenua*, to Nukufetau, it was a direct challenge to her authority for the issues underpinning the claim were matters discussed within the sanctity of the Nukufetau's Indigenous decision making role.

The Nanumea Connection — Inter Island Friendship and Rivalries

The circumstances surrounding Maatia's win had its own touch of *Fenua*. Where Nukufetau's aspirations seemed gone, Nanumea's hope became center stage for the curious mass of people who seemed bent on competing for the earliest coverage of the latest political development, be they factual or sensational rumors. Nanumea and Nukufetau by custom are considered *taugasoa* or friends. But, a friendship with a deeper meaning of traditional honor and respect for each other. It is common practice in Tuvalu and other islands have similar friendship arrangements [Panapa & Fraenkel 2008, p.4]. Nukufetau and Nanumea had that friendship pact prior to colonization when Nukufetau answered Nanumea's request for assistance during a famine.

Could the rumors of the Nanumea move for leadership be related to its traditional ties with Nukufetau? There were so many competing theories covering these uncertain developments. The corridors of the government office buildings were no exemption with its share of gossips. Civil servants are only humans after all and policy provisions for civil servants to remain apolitical are always complicated when *Fenua* is in the equation.

The Funafuti Connection

One fact is not disputed. Mr Kausea Natano of the *Fenua* Funafuti had the majority to form the government with the support of one of Nukufetau's representatives, Lotoala. Nukufetau did not like it but what do you do when this grouping had cordoned themselves in one of the islets until the date for the election of the Prime Minister. Yet, heading to the Vaiaku Lagi Hotel, now the Funafuti Lagoon Hotel, for the election of the Prime Minister, I still heard silent whispers and murmurs of hope among the people who preferred a Maatia government, the same government Nukufetau supports.

The people were hoping on customs worth that Willy, the other Nanumea representative would accept Nanumea's request to support a Maatia Prime Ministership. Whispers of special envoys of Nanumea people who had gone to the islet to speak with Willy were common knowledge.

I wondered though if people were so engrossed in the aspiration for leadership that they failed to accord due respect for the *Fenua* Funafuti and its traditional aspirations, whatever they may have been. After all, most if not all the unsettling events caused by the disconnection in understanding between custom and law, happened on Funafuti, the capital of Tuvalu.

The Making of a Prime Minister

The elections were by secret ballot and presided by the Governor General with the assistance of the Attorney-General and the Secretary to Government. As Attorney-General then, I was directly involved in this election. On hindsight, I found it amusing that all of us in that room may have consciously or unconsciously subscribed to the "sensational rumors publication". We had to count, recount, cross-check and give each other silent questioning looks to make sure we had the correct results. The same results later reflected through the members reaction no doubt supported a theory that they also subscribed to the "sensational rumors publication".

The results swung Maatia's way. I saw Maatia nodding to himself. I wondered if he was silently thanking Willy for who else would have supported him in the circumstances if not the other son of Nanumea. Or, was the nodding a silent thank you to the envoy of Nanumea people rumored to have gone to the islet to seek Willy's support? Whatever the truth is, to most Tuvaluans, Willy had honored his *Fenua*, Nanumea.

The Unmaking of a Prime Minister

Yet, less than three months in office as the new government, the tides turned. Again, Willy was the main actor. To the absolute horror of the government, the hero who had put them in office had helped toppled their leadership by supporting a motion of no confidence in the government for which he was a Cabinet Minister. He later became the Prime Minister. Nanumea still had the position of PM reshuffled between her two representatives. But the reshuffling was brought about by a motion of no confidence tabled by Lotoala of Nukufetau. Rumors about this move had been circulating well before parliament. I believe the government was most disappointed for not having considered traditional grapevine news seriously. After all, the same "publication" was spot on during the successful election of Maatia. I wondered if the current Prime Minister may have been influenced by these experiences when late in 2017 a parliament session was postponed amidst rumors of a motion of no confidence in his government.

Second Chance for Nukufetau

As the saying goes, one person's loss is another's gain. When Maatia's three-month-old government was defeated on 21 December 2010, every candidate was suddenly eligible, again, for the chase to the top job in politics. Would Nukufetau be able to get what it had wanted initially?

Nukufetau tried all it could to facilitate the successful appointment of its representative to the position. Her people wrote letters to Lotoala and to the Governor General, she sent her Chiefs and elders hoping for a dialogue that may support her aspirations, she took to the streets to demand what she believed was hers. As a last resort, she went to Court to seek a confirmation of its Indigenous collectivity understanding, *Fenua's* duty to deal with her subjects and the affairs of the island.

Impact of Ignoring Indigenous Understandings

As the law has it, custom cannot trump the Constitution and one's right as a candidate takes precedent over custom. It was news most unwelcoming to Nukufetau. Application of the law that was contrary to collective understanding resulted in many interesting but unsettling developments. They include but are not limited to the protest March by the Nukufetau community on Funafuti, the birth of a self-proclaimed Nukufetau B clan, the emergence of the notion of *Falesea* or banishment from an island, the sudden dismissal of people from their work not to mention tensions in families divided between the two understandings. The problem was real, and the situation best summed up by Justice Wards in 2012 when he commented in paragraph 20 of the judgement *"At the time of the hearings in this Court, the antagonism of the two groups was tangible."* Nukufetau was not able to maneuver *Fenua or* collective interdependency out of the grips of the law.

Fear the Making of an Unhealthy Mind

Yet, the law does not provide closure to the issue. It facilitates further challenges and disconnections in understandings between law and custom for appeals are yet to be determined. People are left in a state of continuous worry, anxiety and fear of the unknown final verdict. Will the law provide a final verdict? What will *Fenua* offer? For when a judge could feel the tension in the safety of his courtroom and other protected abode, how about the people without such protection. A judge flies into Tuvalu usually for a week or two twice a year for court hearings and returns overseas to his place of residence. However, the people of Tuvalu on Funafuti and the outer islands are subjected, on a daily basis, to the fear of uncertainty. Fear caused by the disconnection in law and Indigenous understanding of *Fenua*. The problem impacts the collective and individual wellbeing.

The Trial — An Eyewitness Account

While Nukufetau's evidence fell short of the legal requirement for removing its parliament representative, the social support for Nukufetau was overwhelming. Whether such social support was right or wrong is beyond the scope of this paper. Rather, the manner in which the islanders showed their support for their *Fenua* Nukufetau was remarkable.

In my years of working in Tuvalu, I never saw a court so packed as it was during these trials. The people in support of Nukufetau's aspirations filled up the surrounding premises of the court room whenever the relevant cases were heard. The courtroom itself could only accommodate the bare minimum of required personnel for any given court hearing.

Open Courtroom

The notion of an open court, in Tuvalu, for this case was uncommon. It was open in every sense of the word, not the Western type with closed doors. Rather, loudspeakers were stationed outside the courtroom to include the number of people the courtroom could not accommodate. It was easy to spot the Nukufetau supporters. They brought the beauty of island festivity to liven up an otherwise boring and tense court atmosphere. The people in their traditional attires of head garlands, white tops and floral *sulu or* wrap around made me wish then if everyone could just get on with friendly traditional dancing, the *fatele*. Yet, I was aware of the eerie reality of the battle between *Fenua* and formal law that was brewing in the courtroom. It was not the time for a *fatele*. The people were still very disturbed by the notion of an individual challenge on the sanctity of *Fenua*. Anything slightly out of the ordinary understanding of collective interdependency, as I was to experience, was a challenge on *Fenua*.

Subpoena

I was subpoenaed as Attorney-General to produce certain documents relevant to these matters. Heading towards the courtroom, about 15 meters in front, was the *palagi* or foreign lawyer handling the Nukufetau and Taukelina cases. He received a standing ovation from the Nukufetau supporters. Tracking five meters behind him was the Secretary to Government who is from Nukufetau. There was no standing ovation. The sad reality of respecting a foreigner over your own. My colleagues and I, a further five meters behind the Secretary to Government found the whole situation hilarious, betting I would get the same cold treatment. I, also of the island of Nukufetau. Spot on, no standing ovation just the usual suspicious look as if we were there to wreck Nukufetau's hopes.

Indigenous Versus Legal Understanding

The Secretary to Government and I were there because of our designations and subpoenaed for matters correctly ruled privileged and not to be used by Counsels. Nevertheless, when you are in a courtroom questioned by Nukufetau's lawyer, for a person of no legal background standing in the crowd, the only respectable conclusion one can draw was that we were working against the wishes of *Fenua*. I understood them though. After all, what layperson understands all the legal complexities that even lawyers struggle with? Reading the statute books is not part of the Indigenous world but a world dictated by statute books, nonetheless. As a Tuvaluan lawyer, I understand that we are all part of *Fenua* and each have our own part to play in maintaining a healthy Tuvalu society.

It is never easy maneuvering your way around custom and law in a society respectful of the law and yet firmly anchored in obedience to collective interdependency. Experience would mentor one to best deal with such a complex situation. So, there I was, the Attorney-General from 8am to 4pm, the official working hours for government, and a walking zombie for the remaining hours. My house was always filled with my people from Nukufetau during the trial periods. They seemed equally divided between *Fenua* and Lotoala. The issues in these cases were a common part of the discussions in our household. Listening to the arguments though, I could tell in most parts that they were clueless about the real legal issues unfolding in court. The tension, even in our household was real and certainly not a healthy situation to be exposed to. I wonder how many other households experienced the same.

A Personal Perspective

Can my mind ever be stripped naked of my Western training and understandings? I do not believe so. I believe that in contemporary Tuvalu there is fertile ground for improving the disconnection in understandings. Both systems are truly entrenched as part of our Tuvaluan life. It is not about elevating one and disregarding the other. It is about extracting the best and salient features of both understandings and constantly weaving a Tuvalaun tapestry conducive to Indigenous living. After all, we are an independent country free to decide what is good for Tuvalu.

As a lawyer, I appreciate the court's decision. But I am also of *Fenua* and sympathize with Nukufetau's concern. As a child of *Fenua*, I do feel for *Fenua* caught in the conundrum of formal laws and Indigenous understanding.

Fenua — Towards an Indigenous Understanding

What caused the disconnection between Indigenous knowledge and law in these matters? Were the laws so complicated that the people could not understand them? Or were they clear but too far removed from Indigenous understanding that Nukufetau could not possibly believe it to be applicable in the context of *Fenua*? There are explanations worth theorizing.

First, everyone in Tuvalu identifies with an island, *Fenua*. However, *Fenua* is not a term one finds easily in the laws of Tuvalu. The laws speak of introduced terms such as *Falekaupule* and *Kaupule* or electorate. While they clarify an aspect of *Fenua*, they continuously reduce *Fenua* to a narrative acceptable to foreign perception but devoid of Indigenous understanding. *Fenua* is commonly understood to be a living thing. One lives with it, breathes it and feels it. This sense of personifying Fenua is better understood through songs where *Fenua* is not just a "piece of land surrounded by water" [Stratford *et al.* 2011, p. 115]. Defining it is not easy though. Every attempt at defining it reduces *Fenua* to something that it isn't as far as Indigenous understanding is concerned.

Secondly, *Fenua* does not distinguish between matters created by statute and those of custom when dealing with its people. Who reads the laws anyway? Aren't laws supposed to be relevant for the subjects it is meant to cover? If so, who needs to read the laws other than carrying on with the norms which should be well reflected in law? But whose norms are the law protecting when the majority of *Fenua* do not agree? Most Tuvaluans identify with an island. They are known by and grounded to an island, *Fenua*. Such grounding comes with DRO. Failing in one's DRO gives rise to various sanctions not the least experienced in the cases discussed and upsets collectivity and individual wellbeing

Thirdly, DRO are guarded with pride. Odd as it may sound for who would want those things? Is one not better off with freedoms and no DRO? Not in Tuvalu, and certainly not for *Fenua*. Where everyone is considered equal, one's pride is in the differences which are selfishly guarded in traditional DRO.

Fourth, matters of importance in Indigenous understanding are not necessarily important in law. In the *Finikaso v Ielemia case*, while the court may have been dealing with the case within the confines of the laws, the real issue for *Fenua* was rather what is this innocent until proven guilty phenomenon? Is it not a straightforward guilty verdict for such leaders? Similarly, in the *Nukufetau v Lotoala* case, *Fenua's* concern was how could the court think that *Fenua* had no control over its parliament representatives, (*Kaupule o Nukufetau v Metia* [2012] TVH 8). How can a representative be of Nukufetau and refuse to meet with the elders of the *Fenua* he is supposed to be representing in parliament? (*Kaupule o Nukufetau v Metia* [2012] TVH 8 Para 10).

Towards a Tuvaluan Bioethics

What has the *Nukufetau v Lotoala* case got to do with bioethics? Well, Like Behren's Africa where "social problems and moral failures are closely linked to human health" so are the spill over effects of the disconnection in custom and law in Tuvalu [Behren 2013, p. 33].

Healthy living in Tuvalu includes the WHO understanding of a "state of complete physical, mental and social wellbeing, and not merely the absence of diseases or infirmity." It includes a harmonious and peaceful mindset [Panapa & Fraenkel 2008]. For Tuvalu, this healthy state of mind is embedded in a healthy *Fenua* and its custom of collectivite interdependency. *Fenua* is core to healthy living and wellbeing in Tuvalu. Custom misplaced is a cause for unhealthy living. Similarly, custom properly utilized is powerful and effective remedy for psychopathologies in diverse cultures [Aina 2017, p. 203].

Ongoing Fear

During my fieldwork in Tuvalu in 2016, the spillover effects of the *Nukufetau v Lotoala* case were still being felt within the communities. Discussions bordering on this event were mostly reduced to whispers. If they were not fidgeting, their eyes would be darting around like a prey conscious of being devoured by a predator at any moment. Something is not right; the tension is still alive. People on either end of the divide remained equally hyper vigilant when discussing the specifics of the case. I wonder if they ever have peace of mind.

While there may be no official report into the health impacts of this particular event in Tuvalu, my observation and discussion with people from 2010 working as the Attorney General of Tuvalu and during my fieldwork in 2016 assures me that Tuvaluans are not immune. They are mentally and socially affected by the impact of the disconnections that arose from misunderstanding customary collectivity or *Fenua*. The continuous disconnection may destroy *Fenua* and collective wellbeing. Importantly, the impacts are exacerbated when the challenges involve the top echelon of Indigenous understanding, *Fenua* and law.

However, in spite of the fear of ongoing challenges, the people still acknowledged the sanctity of *Fenua*. The people I spoke with during my fieldwork in Tuvalu in 2016, people of *Fenua* and the people who experienced the brunt of customary sanctions, remain loyal to *Fenua*. I wondered if they were calculated responses due to fear, distrust or a steadfast acknowledgement of the importance of collective interdependency in their livelihoods amidst all odds. Their respect for *Fenua* was undeterred.

The Way Forward

Bioethics, like many things in the Indigenous world is dominated and influenced by colonial Western values and ethical understandings [Andoh 2011, pp 67–75). *Fenua* is core to the health and wellbeing of Tuvaluans. Where the law weighs heavily in support of individual rights, it impacts negatively on the collective and individual health and wellbeing. I argue that in societies where *Fenua* or collective interdependency remains a strong positive cultural value, there is a pressing need to consider prioritizing it in law.

Similarly, prioritizing it in law does not mean reducing *Fenua's* Indigenous meaning for convenient Western legal drafting. Our tendency to describe *Fenua* within Western legal understandings alters the substance of *Fenua* as understood, felt or lived by the islanders. This continuous shortfall incapacitates *Fenua* because every attempt at a legal fit erodes a part of *Fenua*.

Being described as of collectivity and not individualistic, collective interdependency or DRO, or of collective rights and not individual rights are just parts of *Fenua*. When focus emphasizes only a specific aspect of *Fenua,* it reduces *Fenua* to a creature other than the *Fenua* Indigenous Tuvaluans feel, breath and live. *Fenua* is more than definitions or a piece of land surrounded by waters. *Fenua*, for Tuvalu, can be what WHO considers as that "state of complete physical, mental and social wellbeing". Fenua and law is a health challenge and Tuvaluan bioethics must learn to embrace it.

References

Aina, O. F. (2017). Culture and Mental Health. In O. Omigbodun and F. Oyebode (eds.), *Contemporary Issues in Mental Health Care in Sub-Saharan Africa* (BookBuilders, Nigeria) pp. 203–218.

Andoh, C. (2011). Bioethics and the Challenges to its Growth in Africa, *Open Journal of Philosophy* 1(2): pp. 67–75.

Behren, K. G. (2013). Towards an Indigenous Bioethics, *SAJBL, June 2013*, 6(1) pp. 32–35.

Faaniu, S. (1983). *Tuvalu, A History* (University of the South Pacific, Fiji).

*Census H (2012) Tuvalu 2012.

Government of Tuvalu. (2012). Te Kaniva: Tuvalu Clinate Change Policy 2012.

Isako, T. (1983). Nanumea. In S. Fainuu (ed.), *Tuvalu: A History* (University of the South Pacific, Fiji), pp. 48–58.

Lafai, P. (1983). Nanumaga. In S. Fainuu (ed.), *Tuvalu: A History* (University of the South Pacific, Fiji), pp. 66–71.

Laupepa, K. (1983). Nukufetau. In S. Fainuu (ed.), *Tuvalu: A History* (University of the South Pacific, Fiji), pp. 78–86.

Loomba, A. (2015). *Colonialism/Postcolonialism*, 3rd edition (Routledge, London).

Nations Encyclopedia. (2018). Asia and Oceania. Available from: <http://www.nationsencyclopedia.com/Asia-and-Oceania/index.html>.

Nia, N. (1983). Nuitao. In S. Fainuu (ed.), *Tuvalu: A History* (University of the South Pacific, Fiji), pp. 58–66.

Panapa, P. and Fraenkel, J. (2008). *The Loneliness of the Pro-Government Backbencher and The Precariousness of Simple Majority Rule in Tuvalu* (unpublished thesis). Available from: <https://digitalcollections.anu.edu.au/bitstream/1885/10086/1/Panapa_LonelinessProGovernment2008.pdf>.

Pape, S. (1983). Nui. In S. Fainuu (ed.), *Tuvalu: A History* (University of the South Pacific, Fiji), pp. 71–78.

Stratford, E., Baldacchino, G., McMahon, E., Farbotko, C. and Harwood, A. (2011). Envisioning the Archipelago, *Island Studies Journal* 6(2): pp. 113–130.

Tinilau, V. (1983). Nukulaelae. In Fainuu, S. (ed.), *Tuvalu: A History* (University of the South Pacific, Fiji), pp. 97–102.

United Nations Population Fund, Pacific Sub Regional Office (2012). *Tuvalau National Population and Housing Census. Migration, Urbanization and Youth Monograph.* https://pacific.unfpa.org/sites/default/files/pub-pdf/UNFPA_Tuvalu2012NationalPopulation%26HousingCensusMigration%2CUrbanisationandYouthMonographReportLRv1%28web%29.pdf

https://doi.org/10.1142/9781786348579_0007

CHAPTER SEVEN

Some Implications of Colonialism and Indigenous Cultural Genocide for Healthcare Ethics: Reflections from Northern Ontario, Canada

*Richard Matthews**

"The federal health system is designed to ensure that we remain sick and our people continue to die. First Nations continue to suffer from the shrapnel of a foreign system imposed on us."

James Cutfeet, Chief, Kitchenuhmaykoosib Inninuwug First Nation, Tuesday, February 14 [Porter 2017].

Abstract

Questions of genocide and colonization are barely discussed in bioethics, yet they pose profound moral problems for healthcare ethics. Bioethicists and healthcare workers are accustomed to thinking of healthcare systems as generally beneficent, however much there might be specific flaws in specific institutions, units or individual behaviors. Similarly, theory in health ethics presupposes principalist or virtue-ethical approaches to health ethical issues, and all of them assign

*Associate Professor, Health Advocate and Professionalism Team Lead, Health Sciences and Medicine, Bond University, Gold Coast, Australia.

foundational roles to both law and dialogue in the resolution of complex moral issues. Colonialism, and the genocides which accompany it, challenge these beliefs. For Indigenous peoples, health systems as such are problematic and *a priori*, since these have been, and continue to be, violently imposed upon them. The law that is supposed to function as the final arbiter of moral disputes lies at the heart of the genocides and thus itself can lack legitimacy. That law is the foundation, and that its legitimacy is disputed means that dialogue is superficial rather than real, since any time an Indigenous person objects, and the conflict proves intractable, the means of resolution is violent. Further consequences are that resource allocation decisions become unjustly skewed, beneficence and maleficence shape to dominant settler populations, and health ethical decision-making promotes health inequities rather than eliminating them. In short, racism in health ethical decision-making becomes inescapable. Moreover, given that the destruction of Indigenous identity continues, un-self-critical bioethicists and health workers contribute to that genocide rather than opposing it. This paper describes the problem of genocide — above all that form Lemkin described as 'ethnocide' or 'cultural genocide' and explores some of the practical distortions in ethical decision-making and institutional formation that result.

Introduction

Chief Cutfeet's comment is likely to strike healthcare workers as jarring, challenging as it does altruistic assumptions about the nature of healthcare systems and the people who work in them. His comment forces contemplation of the profound and yet effectively undiscussed implications of colonialism for healthcare ethics. This gap is serious, since all mainstream issues — for example concerning the role of the clinical ethicist, medical assistance in dying, consent and capacity, resource allocation, decision-making in public health — are influenced and distorted by the realities of colonialism. Indeed, the implications are so profound that morally orthodox positions become, paradoxically, the vehicles for racism and injustice.

This essay explores some of the ethical consequences of colonialism and its accompanying genocides. It describes how the routine non-reflexive operations of health law, policy and individual behavior contribute to the social determinants of ill health for these communities.

Healthcare workers, commonly inadvertently, promote the same social determinants of ill health that are causally responsible for the high levels of mortality and morbidity experienced by Indigenous peoples, and do so believing they act for the good. In doing so they promote colonialism and genocide through an uncritical regard for state law, violence and its impacts on contemporary life.

In a colonial society, every aspect of socio-economic life is under-pinned by violence. Colonial societies depend on this to impose a legal, economic, social and cultural regime upon the Indigenous populations they seek to displace. Institutions such as courts, police, child welfare and healthcare institutions possess an ambiguous value at best for colonized populations due to the roles they play in imposing state discipline. Moreover, the institutions served — the legal and economic practices of the colonial state and larger transnational institutions — require the effec-tive marginalization and destruction of independent Indigenous identities, spiritualities, economies and laws since these of necessity conflict with the centralized dominance of a colonial state. For example, private prop-erty laws conflict with Indigenous laws and beliefs about land use and care. In cases of conflict, the police and prison system function to sup-press Indigenous laws and beliefs. The result is that cultural genocide — the elimination of a people as a distinctively identified group — sits at the heart of any colonial society.

This problem is absent from discussions in bioethics. Public health eth-ics is an instructive case.

Public health ethics is concerned with understanding, evaluating and making recommendations on the ethical issues that arise from population health. Above all, it explores the nature and limits of state and institutional intervention in populations [Dawson & Verweij 2008]. Standard issues include the following:

— Balancing communal and individual goods [Bayer & Fairchild 2004];
— The ethical dilemmas connected with epidemics, pandemics, terrorist acts and other states of emergency;
— Identification of appropriate normative frameworks for public health [Gostin 2001; Callahan & Jennings 2002; Goldberg & Patz 2015];

— Appropriateness of political frameworks for public health, including liberal, [Radoilska 2009; Powers *et al.* 2012], libertarian [Menard 2010], capabilities [Venkatapuram 2009; Owen & Cribb 2013], and social justice approaches [Sreenivasan 2009];

— Human rights-based [Mann 1997; Pogge 2005; Nixon & Forman 2008; Ruger 2009] and pluralist accounts of state obligation in balancing liberty, utility and equality in the provision of public health [Selgelid 2009];

— Balancing conflicts of laws, principles and rights in the pursuit of the common good [Kass 2004];

— Paternalist and harm-minimization arguments for state intervention in public health [Holland 2009; Nys 2009];

— The precautionary principle in public health [Weed 2004];

— The relative priority of humanitarian, self-interested and politico-economic considerations in public health [Pogge 2001; Tolchin 2008; Bowleg 2012].

The foregoing offer little in thinking about health ethics with Indigenous peoples. More useful are the following:

— Critiques of statist bias in public health ethics [O'Neil 2002];

— Relational, intersectional and personalist approaches [Baylis *et al.* 2008; Petrini & Gainotti 2008, Bowleg 2012];

— Fears that interventions may impose exploitative conditions on populations or otherwise exacerbate existing inequities [Bhutta 2002; Pinto & Upshur 2009; Braveman 2012; Goldberg 2012];

— Gender and patriarchy in public health [Rogers 2006];

— Economic inequality as a social determinant of health [Hausmann 2009];

— Solidarity in bioethics [Dawson & Jennings 2012];

— The role of medical authority in exacerbating inequities [Wear 2003];

— The distorting effects of racism on healthcare [Gamble 1997].

But none of these explore colonialism or specifically problematize genocide. Of the few healthcare ethics papers that explicitly mention Indigenous health, Michael Marmot addresses the need to account for structural inequalities in Indigenous lives, but challenges neither law, economics nor colonialism. Colonialism and genocide form no part of his framework [Marmot 2012]. And while McNeil *et al.* [2005] mention Indigenous issues in their valuable discussion of disenfranchisement, poverty and racism in healthcare, settler colonialism and Indigenous genocide are oddly invisible. These papers are right to explore privilege and oppression, but a full discussion needs to integrate the social determinants of Indigenous health, colonialism, and ethnocide.

Privilege and Oppression

Privilege and oppression are concerned with unjust distributions of economic, political, social or cultural power. A group is privileged if its members get advantages from the distributions; it is oppressed if its members experience disadvantages and harms. The positioning of the groups is thus dialectically related since the power of the privileged is only possible through the disempowerment of the oppressed, a relationship which is manifest in the health indicators for populations. The improved health experienced by settler populations depends in large part on the ill health inflicted upon Indigenous peoples.

Privilege and oppression are allocated intersectionally, that is across a range of overlapping categories which combine to uniquely determine the conditions of life of each individual. Individuals are members of multiple groups and may be simultaneously privileged in certain categories while oppressed in others. In a capitalist patriarchy, a woman may be oppressed in virtue of being female yet privileged as upper class; in a white privileging capitalist society, someone may be oppressed as transgendered and poor, yet be privileged by whiteness. Comparatively few people only possess privilege markers. The number of those only possessing oppression markers depends on the levels of inequality within a society. In their case, the conditions of life are precarious and the likelihood of survival from day to day is demonstrably low.

All successful colonial societies are defined by Settler privilege and Indigenous oppression since they are founded on the non-consensual seizure of Indigenous lands. Indigenous group members are like other racialized group members, such as (in Canada) Muslims, black Canadians or

Jewish people, in that they have race oppression against them. What distinguishes Indigenous groups is that their culture, economies, spiritual identities, governance systems and land are the primary targets for destruction by settler-colonial states.

Colonialism

Colonialism refers to the complex interplay of institutions of systemic violence — legal, economic, cultural and social — along with the underlying direct violence aimed at entrenching and maintaining Settler supremacy. Through violence, a Settler population aims to impose its economic, political, cultural and spiritual institutions upon one or more Indigenous populations.

Colonialism is a process that is ethnocidal by its nature. A colonial state cannot tolerate the existence of independent legal, spiritual and economic institutions within its borders and thus must suppress and eliminate these. It performs this either through physical genocide, or through the forced assimilation of Indigenous populations to the normative systems and cultural identities of the Settlers [Jones 2006; Short 2010; Palmater 2014; Benvenuto 2015], i.e. through cultural genocide or ethnocide.

The relationship between Settler and Indigenous populations in a colonial state is thereby one of opposition and conflict. It involves ongoing [Vowel 2016] struggles between the Settlers and the Indigenous population(s). One never stops being a Settler even if one's relatives have lived in the region for hundreds of years. The relation is not one of time, but of power and exploitation. An individual is a Settler if they are a beneficiary of the seizures of Indigenous lands. An Indigenous person, in contrast, is someone who understands themselves as tied to their land, heritage, and legal and economic traditions and who resists assimilation to settler-colonial law, economics and culture. Where Indigenous peoples are colonized, a central desire is sovereignty or, to cite Taiaiake Alfred and Jeff Corntassel [2005], the desire to be left alone, un-interfered with by a Settler state. They thereby work to resist assimilation by a colonial socioeconomic regime.

In this regard, colonialism — and ethnocide — is an ongoing process and not merely historical. Colonialism continues for as long as an

Indigenous population resists it. If there is a single Indigenous group asserting its sovereignty and independence, settler entitlement to control land and law remains contested and the violence of colonialism continues.

Colonialism and Indigenous Ill-health

The social determinants of health, for Indigenous people in colonial societies are the causal processes — distal, intermediate and proximal — of colonization. Their impacts — the disproportionately high morbidity and mortality rates experienced by Indigenous populations — are the manifestations of oppression in Indigenous lives.

Colonialism and racism are primary distal determinants of the ill health of Indigenous peoples [Lang 2001; Reading & Wien 2009; Czyzewski 2011]. The land thefts, confinement to reserves, identity damage and resulting poverty drive the intermediate conditions of Indigenous ill health. In Canada, for example, Indigenous people disproportionately lack access to adequate sewage, housing, and water. Education is commonly poorly resourced, and they are forced to live on polluted lands. The result has been poverty imposed across multiple generations. These are the intermediate determinants of the Indigenous suicide crises in Canada, as multiple inquiries have noted [Peoples 1995; Lauwers 2011; 2016; LeFrancois 2016]. The intermediate determinants of ill health in turn drive proximal causal forces such as high rates of mental health and addictions, internal conflict and other forms of suffering inflicted upon many Indigenous communities.

All of these are the product of Canadian law and economic policy and are consequences of the ethnocides. Reserve life, for example, is sharply constrained by the *Indian Act* and the constitution of Canada. Inequitable funding is determined by federal and provincial governments. Indigenous communities have little say and no control over revenue streams. Control over housing, sewage and healthcare is exercised by the Federal government through various federal departments dedicated to Indigenous affairs, as well as through Health Canada. Laws related to resource exploitation, hunting and fishing are set by the federal or provincial governments. All First Nations band council budgets are established by the federal government; the resulting healthcare provided in the nursing stations is inequitable in comparison to equivalent rural non-Indigenous communities [Canada 2015].

Ethnocide and Indigenous Suicide

In this environment, Indigenous suicide rates are disproportionately high. For example, as of 2011, the suicide rate for First Nations males aged 15–24 was five times that of Canadian males generally, and that of Indigenous females from the same demographic was seven times that of Canadian females in general [Lauwers 2011]. There is considerable variation depending on whether the group is First Nations, Metis or Inuit, as well as within spatially contiguous communities and specific Indigenous groups [Kirmayer *et al.* 2007].

Jackie Fletcher, Elder and Commissioner with the Mushkegowuk Council People's Inquiry, correctly ties the suicide rates to Canadian policies of ethnocide or cultural genocide. All of the distal and intermediate determinants of Indigenous suicide are deliberate. They are the result of the partial success of policies aimed at "killing the Indian in the child" and are the ongoing impacts of genocide [Council 2016]. They are instances of what Friedrich Engels called "social murder" [Engels 1845/2010], the infliction of disproportionate levels of physical and mental morbidity and mortality on a population through the application of law and economic policy. Following Engels, the Indigenous suicides are not individual choices, but rather are the consequence of complex group-on-group violence with highly individual impacts. Other problems, such as high levels of mental health and addictions rates, intra-communal violence and higher rates of morbidity and mortality from a range of illnesses, are also the consequence of this violence. In colonial Settler societies, the suicides are a consequence, sometimes explicit, other times tacit, but always clearly preferred and accepted, of the choices of dominant Settler groups to pursue their interests through the dispossession of subordinated Indigenous groups. They result from the deliberate and continuing imposition of sometimes intolerable conditions of life on Indigenous communities.

Given the almost sacred status allocated to law in much bioethics literature, these considerations are significant. If the law is ethnocidal, then what moral position does the healthcare worker occupy? How can dialogue — the *sine qua non* of the successful bioethics consult — be possible? Healthcare institutions and healthcare workers operate within this colonial legal environment and have their choice possibilities sharply constrained by laws that Indigenous peoples often understand — correctly — to be oppressive and racist. To the extent that healthcare workers endorse those laws, they promote that racism, undermine Indigenous access to healthcare, reduce the quality of the care that they do receive, and thereby contribute to the ill health of Indigenous peoples. The following are examples:

Resource Allocation

The reserves are sites of marginalization and confinement of Indigenous peoples. As the Anishnaabe word for reserve — 'Ishgonigan' or 'the left overs' — suggests, they are the sections of land that, at the time of the signing of the treaties, was thought by Canadian authorities to be the least economically valuable. The rest was appropriated as crown lands and allocated for settlement building, as well as in support of resource-based activities such as mining, logging, oil extraction, hunting and fishing, and also for needed logistical supports such as pipelines, railways, roads and power lines. These land seizures continue to determine Indigenous impoverishment as at the time of this writing. Value flows away from the communities, and very little flows back in.

For example, according to the 2015 Auditor General's Report *Access to Health Services for Remote First Nations Communities*, on-reserve healthcare facilities are underfunded in comparison to equivalent healthcare facilities elsewhere in rural Canada. This underfunding extends to all aspects of healthcare, including racial discrimination against 163,000 First Nations children for child welfare support (https://fncaringsociety. com/i-am-witness). What makes this a particular ethical challenge is that the underfunding — which is a major contributor to the social murder of Indigenous people — is a creature of law. Moreover, refusal to change funding levels is tied to cost-benefit calculations that repeatedly conclude that needed clinical and public health improvements are unaffordable and, in any case, outweighed by alternative obligations to fund other "more important" aspects of governance. In short, it is preferable for provincial and federal governments to leave the communities suffering as they do rather than to re-allocate funds from higher priority Settler interests.

Justice and utility discussions which fail to consider the injustice of state colonialism and the requirement for political, economic and social reconciliation inevitably skew funding and perpetuate inequities in healthcare systems and elsewhere. Any decisions are pre-biased by the violent and non-dialogical imposition of settler law and therefore are unjust before any discussions about resource allocation begin, that is unjust *a priori*.

Beneficence

Beneficence is likewise perverted under colonialism. It is common in bioethics to assume the state acts for the best interests of its citizens. Hard decisions get made which, sometimes, work against the interests of specific groups, but in general the assumption is that the state acts for the best. Under colonialism this assumption is incorrect. Rather, beneficence is skewed to the interests of dominant settler-colonial groups and the greater good becomes co-extensive with their needs and desires. Harm-minimization calculations are similarly distorted. The injustices in resource allocation are also examples of distortions in beneficence, since Indigenous groups count less than others in public health decision-making and thus are correspondingly unlikely to be the recipients of beneficence. The result is that, repeatedly, needed public health and clinical interventions are denied because more benefits are routinely adjudged to lie elsewhere.

Epidemics and States of Emergency

The impacts of colonialism, racism and genocide are most apparent in states of emergency. In the last 10 years, various First Nations have declared states of emergency for a wide variety of reasons including:

— youth suicide [Kilpatrick 2016; Neskantaga First Nation 2013; Puxley 2016; Talaga 2017; Woods 2016]

— poor water quality [Bodrug 2015; Eneas 2016; Free Grassy Narrows 2015; Human Rights Watch 2019; Water Canada 2011]
— food [Gerald 2016]
— fentanyl and opioid crises [Jarvie 2016; Mullins 2010; SNTC 2017; AMMSA 2012]
— youth safety [Vis 2017],
— inequities in the provision of health services [Nishnawbe Aski Nation 2016],
— flooding, H1N1 virus [Fitzpatrick 2009]
— and loss of electrical power [Lindell 2011].

This list is not exhaustive.

In some of the cases there was no response of any kind. No government at any level responded to a declaration of state of emergency for suicide by the member communities of Mushkegowuk Council in 2011 [Council 2016]. In other cases, the response is inadequate, such as when the federal government sent body bags instead of vaccines to Wasagamack and God's River First Nations during the H1N1 crisis [CBC News 2009]. Alternatively, federal or provincial governments choose short-term crisis response strategies that leave the causes of the emergencies untouched and thereby guarantee recurrence, as in sending social workers and psychologists to council suicide survivors for a few weeks or months on the rare occasions when an emergency raises sufficient public outrage. Either way, state and private interests act to ensure emergencies are resolved incompetently or not at all.

For many of these emergencies, the response does not meet the nature of the emergency. Far from being exceptional deviations from a norm, the public health emergencies impacting some Indigenous communities are slow-developing, long-standing 'normal' states of affairs. They are the long-standing outcomes of colonial law, policy and economics, i.e. are ethnocidal states of affairs. Hence the causes of the crises are intermediate and distal, and single event crisis response is inadequate. Effective resolution requires reconciliation and supporting long-term political, economic and legal strategies. At root, it requires the elimination of colonialism and transformations in the privilege-oppression relations that currently define colonial states like Canada. In the case of the suicide crises, these are rooted in the racism and economic practices of colonial states, as is clear in the social determinants of ill health for the afflicted communities. Effective anti-suicide strategies need to be multi-layered [Kirmayer *et al.* 2007], long term, and they require a transformation in the socio-economic character of the country.

Respect

Consultation failures and breakdowns are common in Indigenous health. Indigenous communities either do not get asked about their interests or needs, or they are consulted in a perfunctory manner. Under these conditions, the identity and autonomy of Indigenous peoples is de facto irrelevant, and the result is decision-making that is either of no benefit to Indigenous communities, or positively harmful. Not only are communities ignored, but the specific health knowledge and expertise of Indigenous healers and law keepers is routinely dismissed in spite of the extensive and intimate ecological and public knowledge that they possess.

An example from northern Ontario is of value in considering the impacts of ignoring Indigenous voice. A great deal is known about suicide prevention, and about the causes of suicides for both Settler and Indigenous populations. Successful anti-suicide programs have been instituted that are relatively cheap. Health Canada had established a successful program at Wakapeka First Nation and then cut its funding in 2016. In July of that year, the band council warned Health Canada that a suicide cluster was developing and that the funding needed to be restored. The call was rejected. Six months later, in January 2017, two 12-year-old girls died, followed by another death by suicide in May 2017 of another child, along with an attempted suicide by an additional teenager. The communities were not consulted on their preferences about funding and have been dispossessed of control over the land and resources which would allow them to shape their own healthcare.

'What Should We Do?'

Commonly, well-intentioned people will ask this question. However, the question itself misunderstands the problems. Above all it is inadequate because it leaves the position, power and character of the questioner untouched. Those who pose this question fail to see how their power to act lies at the heart of Indigenous ill health. As long as the oppressive power relations remain intact, little happens other than the occasional band-aid. This is shown, for example, in more progressively oriented people who defend resource transfers — for example, extra funding for nursing stations, mental health and addictions programs, and the like — without challenging their own right unilaterally to decide on policy and resource allocation. The inequitable allocations are only possible given the suppression of the community's capacity to decide since the communities obviously would try to adequately fund their own healthcare. The actual obligations are about giving up power, i.e. about sharing or otherwise redistributing it and reducing control over resources.

Who Should We Be?

The central moral challenge is virtue-ethical. Healthcare workers and settlers generally need to stop asking what we should do and first ask who we, as a society, need to become. Successful efforts at improving Indigenous health and lives depend upon changing the relationship between settler and Indigenous populations and are thus about transforming relations of power. Consequently, transformation of the identity and behaviors of healthcare worker, the structures of healthcare systems and the colonial society from which they benefit is foundational. This means above all the elimination of systemic, epistemic and interpersonal racism as they manifest in the mundane operations of healthcare. The social determinants of Indigenous ill health implicate all of us — including health institutions and healthcare workers — as causally responsible for the high rates of mortality and morbidity in Indigenous communities. Thinking about solutions can make sense once these facts about colonialism and racism are recognized. But it means that we take seriously Chief Cutfeet's assertions about the designed harms of colonial healthcare systems. It also means that health ethics has to prioritize Indigenous sovereignty and support meaningful justice-based compensation for past and present wrongs.

The Moral Primacy of Advocacy

Work and research in clinical and public health matter for everyone. However, under conditions of colonialism and racism, any health efforts are limited or without effect because of the consequences of privilege and oppression in settler colonial societies. Under such conditions, the most effective moral work is reflective, anti-colonial social justice advocacy. Such advocacy has to focus on reconciliation and redistributions of power. Given that colonialism is fundamentally concerned with the seizure of control over resources, the real challenge is to shift the socio-economic situation so that state and corporate authorities no longer exercise unilateral decision-making power over land, resources, education and health philosophy.

The ethical healthcare worker, in working with Indigenous populations, has to advocate for land restorations, transformations in law and, ultimately, for decolonization — that is a transformation in the relationship between the colonial state and the Indigenous populations which it oppresses. If it is to be adequate to the ethical demands of health ethics with Indigenous peoples, bioethics has to incorporate decolonization theory. This means supporting land transfers, land rights, and reparations/reconciliation payments along with the provision of culturally safe health expertise — where requested — in support of healthcare interventions.

Conclusion

Some excellent work in Indigenous health is being done by many different players within healthcare. But unless the distal and intermediate determinants of ill health are transformed, the possibilities of improvement are limited and vulnerable to abandonment whenever other colonial interests dictate. Such improvements require transformations in the politics, law and economic practices of colonial societies. There may be alleviation of harm in some cases, but little likelihood of broad improvements in health. Eliminating the suicide crises and other social problems is highly unlikely without changes in macro-economic resource allocation, large scale decontamination of Indigenous lands, improvements in housing, sanitation, access to recreation, cultural and spiritual opportunities, access to and control of land, and recognition of Indigenous sovereignty and law. Colonial states like Canada continue to suppress these. This is the environment within which healthcare workers function, and we have to negotiate it ethically as best as we can. In this regard, the ethically minded healthcare worker has to work against the lines of colonial force. This is risky, since it can place us in opposition to interests which include our employers. It also means that the healthcare worker may have to act in ways which are risky, provide no reward, and which may be punished by the incentive structures of their employers and institutions. Healthcare workers are not entitled.

Acknowledgements

I would like to thank Ms. Emma Woodley for her careful editorial work. I would also like to acknowledge the wisdom and generosity of Anishnaabe healers Tom Chisel and Ralph Johnson, without whom my understanding of these problems would be shallow.

References

Alfred, T. and Corntassel, J. (2005). Being Indigenous: Resurgences against Contemporary Colonialism, *Government and Opposition* 40(4): pp. 597–614.

AMMSA (2012). Cat Lake First Nation Declares State of Emergency. Available from: <https://ammsa.com/publications/windspeaker/state-emergency-was-declared-cat-lake-first-nation-northern-ontario>.

Bayer, R. and Fairchild, A. L. (2004). The Genesis of Public Health Ethics, *Bioethics* 18(6): pp. 473–492.

Baylis, F., Kenny, N. P. and Sherwin, S. (2008). A Relational Account of Public Health Ethics, *Public Health Ethics* 1(3): pp. 196–209.

Benvenuto, J. (2015). What Does Genocide Produce? The Semantic Field of Genocide, Cultural Genocide, and Ethnocide in Indigenous Rights Discourse, *Genocide Studies and Prevention: An International Journal* 9(2): pp. 26–40.

Bhutta, Z. A. (2002). Ethics in International Health Research: A Perspective from the Developing World, *Bulletin of the World Health Organization* 80(2): pp. 114–120.

Bodrug, C. (2015). Dokis First Nation Under State of Emergency, *My Northbay Now*, June 11. Available from: <http://www.mynorthbaynow.com/6504/dokis-first-nation-under-state-of-emergency/>.

Bowleg, L. (2012). The Problem with the Phrase 'Women and Minorities': Intersectionality — an Important Theoretical Framework for Public Health, *American Journal of Public Health* 102(7): pp. 1267–1273.

Braveman, P. (2012). We are Failing on Health Equity Because we are Failing on Equity, *Australian and New Zealand Journal of Public Health* 36(6): p. 515.

Callahan, D. and Jennings, B. (2002). Ethics and Public Health: Forging a Strong Relationship, *American Journal of Public Health* 92(2): pp. 169–176.

Canada, O. o. t. A. G. o. (2015). *Access to Health Services for Remote First Nations Communities*. Office of the Auditor General of Canada.

CBC News (2009). Ottawa Sends Body Bags to Manitoba Reserves, September 16. Available from: <https://www.cbc.ca/news/canada/manitoba/ottawa-sends-body-bags-to-manitoba-reserves-1.844427>.

Council (2016). Nobody wants to die — they want the pain to stop. The People's Inquiry into our Suicide Pandemic. Mushkegowuk Tribal Council.

Czyzewski, K. (2011). Colonialism as a Broader Social Determinant of Health, *The International Indigenous Policy Journal* 2(1): pp. 1–14.

Dawson, A. and Jennings, B. (2012). The Place of Solidarity in Public Health Ethics, *Public Health Reviews* 34(1): pp. 65–79.

Dawson, A. and Verweij, M. (2008). Public Health Ethics: A Manifesto, *Public Health Ethics* 1(1): pp. 1–2.

Eneas, B. (2016). Update: Muskoday First Nation Continues State of Emergency, *PA Now*, July 28. Available from: <http://panow.com/article/582407/muskoday-first-nation-declares-state-emergency>.

Engels, F. (1845/2010). *Condition of the Working Class in England* (Panther Edition, Moscow).

Fitzpatrick, M. (2009). First Nations Call for Swine Flu State of Emergency, *Montreal Gazette*, June 25. Available from: <http://www.montrealgazette.com/business/first+nations+calls+swine+state+emergency/1729124/story.html>.

Free Grassy Narrows (2015). Grassy Narrows First Nations Declares Emergency Over Bad Water, August 27. Available from: <https://freegrassy.net/grassy-narrows-declares-state-of-emergency-over-unsafe-drinking-water/>.

Gamble, V. N. (1997). Under the Shadow of Tuskegee: African Americans and Healthcare, *American Journal of Public Health* 87(11): pp. 1773–1777.

Gerald, J. (2016). Why Aren't Conditions of Life for First Peoples a National Emergency? *Canadian Dimension*, April 6. Available from: <https://canadiandimension.com/articles/view/why-arent-conditions-of-life-for-first-peoples-a-national-emergency>.

Goldberg, D. S. (2012). Social Justice, Health Inequities, and Methodological Individualism in US Health Promotion, *Public Health Ethics* 5(2): pp. 104–115.

Goldberg, T. L. and Patz, J. A. (2015). The Need for a Global Health Ethic, *The Lancet* 386 (November 14): pp. 37–39.

Gostin, L. O. (2001). Public Health, Ethics, and Human Rights: A Tribute to the Late Jonathan Mann, *Journal of Law, Medicine and Ethics* 29: pp. 121–130.

Hausmann, D. M. (2009). Benevolence, Justice, Well-Being and the Health Gradient, *Public Health Ethic*s 2(3): pp. 235–243.

Holland, S. (2009). Public Health Paternalism — a Response to Nys, *Public Health Ethics* 2(3): pp. 285–293.

Human Rights Watch (2019). Canada: Blind to First Nations Water Crisis. Available from: <https://www.hrw.org/news/2019/10/02/canada-blind-eye-first-nation-water-crisis>.

Jarvie, M. (2016). Stoney First Nations Facing Prescription Addiction Crisis, *Calgary Herald*, July 25. Available from: <http://calgaryherald.com/news/local-news/stoney-first-nations-facing-prescription-addiction-crisis>.

Jones, A. (2006). *Genocide: A Comprehensive Introduction* (Taylor & Francis, London).

Kass, N. (2004). Public Health Ethics: From Foundations and Frameworks to Justice and Global Public Health, *Journal of Law, Medicine and Ethics* 32(2): pp. 232–242.

Kilpatrick, S. (2016). Attawapiskat: Four Things to Help Understand the Suicide Crisis, *The Globe and Mail*, April 11. Available from: <https://www.theglobeandmail.com/news/national/attawapiskat-four-things-to-help-understand-the-suicidecrisis/article29583059/>.

Kirmayer, L. J., Brass, G. M., Holton, T. Paul, K., Simpson, C. and Tait, C. (2007). *Suicide among Aboriginal People in Canada* (The Aboriginal Healing Foundation, Ottawa).

Lang, T. (2001). Public Health and Colonialism: A New or Old Problem? *Journal of Epidemiology & Community Health* 55: pp. 162–163.

Lauwers, B. (2011). The Office of the Chief Coroner's Death Review of the Youth Suicides at the Pikangikum First Nation 2006–2008. Office of the Chief Coroner for Ontario, Toronto.

LeFrancois, B. (2016). *Report of Inquest* — Act Respecting the Determination of the Causes and Circumstances of Death for the Protection of Human Life Concerning the Deaths of Charles Junior Gregoire-Vollant, Marie-Marthe Gregoire, Alicia Grace Sandy, Celine Michel-Rock, Nadeige Guanish. Bureau du Coroner Quebec, Quebec City.

Lindell, R. (2011). Two Quebec First Nation Communities Declare States of Emergency, *The Province*, December 5. Available from: <http://www.theprovince.com/business/quebec+first+nations+communities+declare+states+emergency/5819708/story.html>.

Mann, J. (1997). Medicine and Public Health, Ethics and Human Rights, *The Hastings Centre Report* 27(3): pp. 6–13.

Marmot, M. (2012). Health Equity: The Challenge, *Australian and New Zealand Journal of Public Health* 36(6): p. 513.

McNeil, P. M., Macklin, R., Wasunna, A. and Komaseroff, P. A. (2005). An Expanding Vista: Bioethics from Public Health, Indigenous and Feminist Perspectives, *Medical Journal of Australia* 183(1): pp. 8–9.

Menard, J. F. (2010). A 'Nudge' For Public Health Ethics: Libertarian Paternalism as a Framework for Ethical Analysis of Public Health Interventions? *Public Health Ethics* 3(3): pp. 229–238.

Mullins, K. (2010). Canadian First Nation Community Declares State of Emergency, *Digital Journal*, October 22. Available from: http://www.digitaljournal.com/article/299264

Neskantaga First Nation (2013). Neskantaga First Nation Declares a State-of-Emergency Following Recent Suicides, News Release, April 17. Available from: <http://www.nan.on.ca/upload/documents/nr-neskantaga---april2013.pdf>.

Nishnawbe Aski Nation (2016). *Declaration of a Health and Public Health Emergency in Nishnawbe Aski Nation (NAN) Territory and the Sioux Lookout Region*. Available from: <http://www.nan.on.ca/upload/documents/comms-2016-02-24declaration-health-emerg.pdf>.

Nixon, S. and Forman, L. (2008). Exploring Synergies between Human Rights and Public Health Ethics: A Whole Greater than the Sum of its Parts, *BMC International Health and Human Rights* 8(2): pp. 1–9.

Nys, T. (2009). Public Health Paternalism: Continuing the Dialogue, *Public Health Ethics* 2(3): pp. 294–298.

O'Neil, O. (2002). Public Health or Clinical Ethics: Thinking Beyond Borders, *Ethics and International Affairs* 16(2): pp. 35–45.

Owens, J. and Cribb, A. (2013). Beyond Autonomy and Individualism: Understanding Autonomy for Public Health Ethics, *Public Health Reviews* 6(3): pp. 262–271.

Palmater, P. (2014). Genocide, Indian Policy, and Legislated Elimination of Indians in Canada, *Aboriginal Policy Studies* 3(3): pp. 27–54.

Peoples, R. C. o. A. (1995). *Choosing Life: Special Report on Suicide among Aboriginal People*. Ministry of Supply and Services, Ottawa, Canada.

Petrini, C. and Gainotti, S. (2008). A Personalist Approach to Public Health Ethics, *Bulletin of the World Health Organization* 86(8): pp. 624–629.

Pinto, A. D. and Upshur, R. E. G. (2009). Global Health Ethics for Students, *Developing World Bioethics* 9: pp. 1–10.

Porter, J. (2017). Ontario First Nation 'Reeling' After Suicide of 11-Year-Old Girl, *CBC News*, Feb 15. Available from: <http://www.cbc.ca/news/canada/thunder-bay/11-year-old-suicide-1.3982955?cmp=abfb>.

Pogge, T. (2001). Priorities of Global Justice, *Metaphilosophy* 32(1/2): pp. 6–24.

Pogge, T. (2005). Human Rights and Global Health: A Research Program, *Metaphilosophy* 36(1/2): pp. 182–209.

Powers, M., Faden, R. and Saghai, Y. (2012). Liberty, Mill and the Framework of Public Health Ethics, *Public Health Ethics* 5(1): pp. 6–15.

Puxley, C. (2016) Manitoba First Nation Declares State of Emergency Over Suicide Epidemic, *The Globe and Mail*, March 9. Available from: <https://www.theglobeandmail.com/news/national/manitoba-first-nation-declares-state-of-emergency-over-suicide-epidemic/article29113402/>.

Radoilska, L. (2009). Public Health Ethics and Liberalism, *Public Health Ethics* 2(2): pp. 135–145.

Reading, C. and Wien, F. (2009). *Health Inequalities and Social Determinants of Aboriginal Peoples' Health* (National Collaborating Centre for Aboriginal Health, Prince George, BC).

Rogers, W. A. (2006). Feminism and Public Health Ethics, *Journal of Medical Ethics* 32: pp. 351–354.

Ruger, J. P. (2009). Global Health Justice, *Public Health Ethics* 2(3): pp. 261–275.

Selgelid, M. (2009). A Moderate Pluralist Approach to Public Health Policy and Ethics, *Public Health Ethics* 2(2): pp. 195–205.

Short, D. (2010). Cultural Genocide and Indigenous Peoples: A Sociological Approach, *The International Journal of Human Rights* 14(6): pp. 833–848.

SNTC (2017). Secwepemc Elders and Chiefs Declare State of Emergency over Fentanyl Crisis, Press Release, March 9. Available from: <http://shuswapnation. org/secwepemc-elders-and-chiefs-declare-state-of-emergency-over-fentanyl-crisis/>.

Sreenivasan, G. (2009). Ethics and Epidemiology: Residual Health Inequalities, *Public Health Ethics* 2(3): pp. 244–249.

Talaga, T. (2017). Northern Ontario First Nation Declares State of Emergency on Youth Suicides, *The Star*, June 22. Available from: <https://www.thestar. com/news/canada/2017/06/22/northern-ontario-first-nation-declares-state-of-emergency-on-youth-suicides.html>.

Tolchin, B. (2008). Human Rights and the Requirement for International Medical Aid, *Developing World Bioethics* 8(2): pp. 151–158.

Venkatapuram, S. (2009). A Bird's Eye View: Two Topics at the Intersection of Social Determinants of Health and Social Justice Philosophy, *Public Health Ethics* 2(3): pp. 224–234.

Vis, M. (2017). Nan Calls for State of Emergency, *TB Newswatch*, July 7. Available from: <https://www.tbnewswatch.com/local-news/nan-calls-for-state-of-emergency-665065>.

Vowel, C. (2016). Indigenous Writes: A Guide to First Nations, Metis & Inuit Issues in Canada (Highwater Press, Winnipeg, MB).

Water Canada (2011). State of Emergency in Pikangikum. Available from: <http://watercanada.net/2011/state-of-emergency-in-pikangikum/>.

Wear, D. (2003). Insurgent Multiculturalism: Rethinking How and Why We Teach Culture in Medical Education, *Academic Medicine* 78(6): pp. 549–554.

Weed, D. (2004). Precaution, Prevention, and Public Health Ethics, *Journal of Medicine and Philosophy* 29(3): pp. 313–332.

Woods, A. (2016). 28 States of Emergency are In Effect in these Ontario Communities, *The Hamilton Spectator*, April 25. Available from: <https:// www.thespec.com/news/ontario/2016/04/25/28-states-of-emergency-are-in-effect-in-these-ontario-communities.html

https://doi.org/10.1142/9781786348579_0008

CHAPTER EIGHT

Indigenous Research Contexts and Social Equipoise: Comparative Observations on Tensions in Ethical Oversight in New Zealand, Australia and Canada

*Barry Smith**

Abstract

While the idea of Indigenous ethics frameworks has gained currency over the past twenty years, with the appearance of documents and guidelines that reflect the specific issues of Indigenous populations, the extent to which identified anxieties are taken seriously in the ethics space remains a concern. Using global examples, but especially focusing on Australia, Canada and New Zealand, this chapter examines the tensions between 'mainstream' health ethics and the incorporation of Indigenous principles. The account here of these principles are generated from cultural factors and from observations regarding the disparity in health outcomes and health status between Indigenous populations and other segments in society. Also considered is the tendency to focus attention on the 'soft' issues of culture as opposed to looking at ethics evaluation in terms of 'social equipoise'. Social equipoise refers to the process where the research question, design and research processes have the

---

*We have co-dedicated this book to Barry Smith, who passed away in February 2019. He was active in the Lakes District Health Board, Rotorua, New Zealand.

potential to deliver benefits and serve the best interests of Indigenous peoples without the imposition of undue risk. This process as a way forward in thinking about ethical review of research is discussed.

Introduction

The characteristics of the social and economic conditions experienced by Māori historically and presently can be laid firmly at the door of the colonization process involving the settler peoples from Europe in the eighteenth and nineteenth centuries [Smith 1999]. The impacts of this contact have been long lived and persistent, with Māori exhibiting statistics indicating social and economic inequality across a wide range of indicators. Data on incarceration [Newbold 2007], health status, mental health and life expectation [Reid & Robson 2006] glaringly demonstrate the degree to which disparity exists between Indigenous and 'mainstream' populations. No matter how governments have claimed the existence of policies aimed at 'closing the gap', the achievements have been disappointing.

Of course, in the New Zealand context, the 1970s onwards has seen a resurgence and increased vitality from within Māori around the need to ensure the resilience of their own social structures, processes and changing cultural identity [Poata-Smith 2013] and to promote an understanding of these elements globally. Much of this renaissance has been focused on the protection and maintenance of cultural capital in respect of language and cultural processes, along with the need to recognize the value attached to 'parallel' processes with regard to areas such as education, health and justice. Concurrently, energy has been expended on the goal of effectively working through past injustices with regard to land and other economic resources experienced from the time of first settler contact [Cunningham & Stanley 2003]. 'Righting past wrongs' has been reflected in much of the political dialogue taking place within government, especially around the prioritization and funding of research. It is in this context that the conversation around the distribution of harms and benefits becomes immediately relevant. Other countries such as Australia and Canada have and are facing similar issues and the comparisons in the attempts to deal with cultural difference in the ethics space will be outlined below.

Social Equipoise

Equipoise is a commonly met term in ethical review. Thus, when comparing two drugs in a clinical trial, the only acceptable assumption is that the researcher does not know whether one drug has greater efficacy than the other. As applied in an ethical evaluation, equipoise implies that there is a balance of risks and benefits in respect of each intervention and that they are hence 'equally poised'. On the other hand, if drug A is known to be more effective in treating a condition than drug B then a trial is not exhibiting equipoise. To make the comparison under these conditions would knowingly put patients (unnecessarily) at risk without their knowledge and, more importantly, when there is no need to carry out the research in the first place. Thus, in medical research, equipoise is more formerly viewed as occurring when there exists ignorance around the relative impact of the agent or the intervention being tested against a known agent or intervention, often approached via a methodology referred to as RCT or randomized controlled trial.

In this chapter I descriptively explore an extension of the term equipoise by applying it to ethics oversight and the way in which the relationships between the research process and Indigenous communities are shaped and controlled. The discussion looks at the concept of 'social equipoise', using it as a label to mark out situations where there exists an inequitable distribution of risks and benefits in a piece of research associated with *different categories of participant* as opposed to different interventions. The term social equipoise has been used in a variety of writings and disciplines from social history and law [e.g. Merges 2011] to education [e.g. Theobald 1993]. In these contexts, the common meaning is associated with, and is taken to depict, a state of balance that is imbued with a state of 'natural' harmony. However, used in this sense, a state of social equipoise is not how we would describe the current levels of health status and well-being of Māori in New Zealand. The easily sustained thesis is that Māori, due to the impact of a colonial past, are more likely to be exposed to greater risks and to gain fewer benefits from health research whether measured in terms of individual participants or the communities from which these participants are drawn.

The Research Environment

The nature of the research process has been viewed as being central to, and indicative of, the continuation of injustice. As Smith [1999, p. 49] notes:

> Western research brings to bear on any study of Indigenous peoples, a cultural orientation, a set of values, a different conceptualization of such things as time, space, and subjectivity, different and competing theories of knowledge, highly specialized forms of language, and structures of power.

As the social and health sciences have matured, research that more accurately describes and understands the root 'causes' of these disparities has also gathered momentum. One consequence of this has been the (slower than desirable) realization that better research and more 'useful' and 'relevant' results and outcomes has the potential to support a reduction in differences in social and health status, thus moving towards a state of social equipoise.

The need to understand better the nature of the impacts of colonization and to find ways to locate and frame effective solutions to these impacts has effectively promoted an emphasis on the value of research and research capacity. However, given the shape of the history of the interaction between Indigenous and settler populations globally through to the present, it has become apparent that the nature of the relationship between Indigenous and research communities can either, conform to a historical 'colonized' format, or take on an arrangement that reflects the reconstituting and redefining of the relationship promoted by this renaissance.

Health Disparity

The willingness to focus on health disparities, in spite of the conspicuous absence of this issue from broad political debate in New Zealand, can be illustrated in a sample of research efforts over the past decade. In a joint Ministry of Health and University of Otago publication *Decades of Disparity III* [2006, p. xii], the conclusion is reached that:

> Socioeconomic position and ethnicity exert both joint and independent effects on mortality through multiple pathways, and both must be addressed through health and social policy setting....

Examples of disparity also take on a locational flavor with Pearce *et al.* [2008] concluding that decreases in mortality have not been consistent for all causes of mortality or all regions reflected in an increase in geographical inequalities over time. Compounding this effect is the work of Rumball-Smith [2009] which suggests that there is some evidence for differences in the quality of in-hospital care for Māori in New Zealand. On a global level, Wilkinson and Pickett [2009] generate data to show that there is a relationship between the degree of social inequality and health and social outcomes. In terms of this paper, the critical point is that, of the 23 developed countries chosen, New Zealand was ranked as fifth from bottom with Australia and, less so Canada, exhibiting serious degrees of income inequality.

In research terms, Māori and non-Māori cannot be grouped as equally poised in terms of life chances, with the resulting argument justifying a greater focus on research with Māori supported in terms of the need to equalize distributive justice. However, when reviewing annual reports of the Health and Disability Ethics Committees (HDECs), particularly in first decade of the 2000s, the overwhelming sense contained in the comments of Māori representatives on these committees is that there was an obvious disconnect in both the research and ethics evaluation contexts as far as giving a much needed emphasis to the issue of health disparity between Māori and non-Māori. Solving this inequality was perceived by these observers to not be uppermost in researchers' minds with Māori members of ethics committee, as just noted, often recording their view that the level of responsibility associated with reducing health disparity within the health research space is sorely inadequate. For example, these members point to:

> **inadequacies** seen in the way applicants engage with Māori" (Multi-region Ethics Committee).
>   ...applicants failing to provide basic information about the relevance of studies for Māori (Northern X Ethics Committee).

and note that:

> **Confusion** remains over the cultural component of the ethics application, and this creates a wide variety of responses that are often **less than**

adequate to address cultural impacts of the research on Māori... (Northern Y Ethics Committee).

The committee has received a range of protocols that still **have not considered Māori consultation**. (Central Ethics Committee).

Another issue highlighted throughout the year was the apparent **inconsistency** in the answers to questions completed by researchers. (Upper South B Ethics Committee).

Devaluing Indigenous Constructs and Paradigms

In addition, there are a number of presumptions that researchers seemingly bring to the ethical review process that devalue and undermine the credibility of Māori constructs and frameworks. The most critical of these include the view that: (1) key ethical concepts are independent of culture; (2) 'Western' concepts and explanatory frameworks are universally applicable; (3) Māori constructs (referred to as 'matauranga Māori') are inadequate for the systematic investigation of (health) phenomena and, Māori in any case use 'Western' concepts to understand their world and, finally, (4) there exists a 'pan-Māori' world view. This set of propositions might be translated into the vernacular as — "if we get the cultural bits right then everything else should be ok too remembering that, while culture is your domain, science is ours". But, in terms of expanding our understanding regarding the determinants of health inequality, the attitude that there is little value to be gained from pursuing engagement over and above the concern with culture, and so ignoring the development of relationships at a paradigmatic and intellectual level, remains problematic.

The Correlates of Principalism and Positivism in Cross-cultural Contexts

The problem just described also resides within the process of ethics review itself where the elements of Principalism, which underlie much ethics evaluation, and the presence of positivism which permeates health research, come together to support a set of attitudes and practices that is antithetical to effective engagement with the health issues of Māori. A reading of the early Health and Disability Ethics Committees (HDECs) annual reports noted above, and the results of a survey on attitudes to Māori consultation and responsibilities around the Treaty of Waitangi carried out on applicants to the HDECs, together with observations made of HDEC meetings [see Tolich & Smith 2015 esp. Ch 9] indicate that a number of key factors are at play here. First, the attitudes expressed, and behaviors exhibited by researchers suggest the presence of a degree of cultural tokenism and an absence of 'deep' cultural understanding. The implications of this are the presence of a superficial contextual position coupled with a focus on opportunities to access data rather than to promote a focus on the potential realization of health benefits. From a global Indigenous perspective, this fuels the view that this amalgam of factors promotes and leads to the use of 'inappropriate' methodologies that frequently produce irrelevant and 'un-meaningful' results! This reduces the motivation for effective research translation thus diminishing the opportunities for the uptake and beneficial application of research findings. What all of this indicates is a clear requirement for change in the perspectives and behaviors of both the health research and ethics communities.

Changes to the Nature of Health Research

Amongst the desired modifications argued for would be that, within the research community, greater cognizance be taken of models of social justice and resource allocation together with the purposeful targeting and the balancing of emphasis between health improvement and health maintenance — that is between targeted intervention and universal health care. Of course, this will be more likely to occur if there were greater cultural empathy and understanding on the part of researchers and research agencies and a more 'pragmatic' approach to discussions and debates around the links between competing cultural paradigms and health outcomes. In other words, what must be more visible is the intention to discuss not just cultural matters but also the research project and its primary goals and questions and that this occurs early in the life of the project or program. Moreover, this in turn is more likely to occur in a context defined by the greater presence of research-community links and partnerships. But what of the part played by the ethics review in the route to social equipoise?

Changes to the Approach to Ethics Review

In New Zealand, the way in which ethics committees have dealt with the issue of their responsibilities under the Treaty of Waitangi and with the quality of engagement with Māori by the research community shows both complexity and inconsistency. However, a reasonable and appropriate position in the context of the degree of New Zealand's social and economic inequality is that the ability to support a focus on health disparity and the equitable distribution of risk and benefits from health research will only occur if there is a move away from both ethical and methodological 'fetishism' — that is, an adherence to structures and processes that obviously do not work for Māori. Ideally, this would be coupled with a willingness to understand the value of other research methodologies which has largely been prevented by the obvious hegemony of the biomedical model as the ever-present measure of research credibility within the health research environment. Such a position would also be supported by an acceptance of the existence of intrinsic value in Māori constructs and paradigms[1] together with a greater understanding within the ethics review process of aspects of both diversity and the dependency thesis reflected in the form of a mild version of ethical and cultural relativism. Certainly, the availability of culturally specific but generic guidelines such as *Te Ara Tika: Guidelines for Māori Research Ethics* [Hudson *et al.* 2010] and the more recent and specific *Te Mata Ira: Guidelines for Genetic Research with Māori* [Hudson *et al.* 2016], both of which suggest firmer definitions and clearer sanctions around the quality and substantiveness of Indigenous engagement, should assist in the achieving a more equitable distribution of harms and benefits in the health research space.

The Place of Indigenous Ethics Frameworks

Ultimately, the value of such documents and the conversations they engender is (simply) that they provide a realization that there exist

---

[1]Tied into research funding processes run by the Ministry of Business, Innovation and Employment (MBIE), this sentiment is captured by a policy statement referred to as *Vision Matauranga*.

different perspectives and different questions that become important when there is a genuine desire to more fully understand Indigenous contexts in both a societal and research sense. The core proposition of this chapter is that the commentary above, and the modifications contained therein, will lead to a fuller understanding of factors contributing to health status and health inequalities, by way of a greater emphasis on 'social' versus 'biomedical' models together with an improved insight into Indigenous paradigms around health and well-being resulting in a raising of the degree of cultural credibility with Māori participants and their communities. Alongside of this should result better cultural insight into the *application* of ethical concepts around matters such as collective versus individual consent, confidentiality and the ethics of dignity and maintenance of 'mana' (status and influence) referred to as 'non-physiological harm'. These ingredients and the raised credibility will provide the potential for greater uptake and application of research findings and a greater opportunity to give consideration to the discussion of distributive justice. All of these elements are precursors to a situation where there exists a more equitable distribution of harms and benefits between different categories of participants and thus the state of social equipoise. A consideration of what can be learnt from other jurisdictions that have 'first nations' populations, namely Australia and Canada, may be useful.

Hands Across the Water: Comparative Efforts in the Indigenous Research Space

In Australia, government policy concerning Indigenous consultation has developed step-by-step procedures that set out how researchers are to engage with Indigenous communities. On the other hand, research not related to Indigenous communities is exempt from mandatory Indigenous consultation. By comparison, in New Zealand, everything is defended with all researchers implicitly bound by the Treaty of Waitangi and with this an obligation to consult with Māori under a variety of operational standards and guidelines. Yet unlike Australia, New Zealand researchers as yet have no unequivocal guidelines that support and facilitate this form of consultation. This position results from the unforeseen consequences of

Māori importantly and rightfully championing essentialist kaupapa Māori research noting that this research is not only done with Māori, it is done *by* Māori. Non-Māori researchers are thus seen to have a limited role in this methodological context [Smith 1999]. Australia has no equivalent; its guidelines are inclusive not exclusive within this restricted scope of engagement. The interesting irony is that kaupapa Māori research's (informal) axiom around the perceived exclusion of non-Māori researchers has resulted in a situation where specific guidelines for Indigenous consultation for non-Māori has languished, resulting in what has been described as paralyzed best practice [see Tolich 2002].

It is thus instructive to describe the nature of this irony in New Zealand contrasting it against the Australian and Canadian regulations and perspectives on Indigenous[2] research. Such a description of course acknowledges the background similarities between the white settler colonies in both countries and explains the value of clear ethical guidelines for those researching Indigenous communities. Of particular interest though is the divergence in the types of Indigenous research *requiring mandatory consultation* across the three countries which, in turn, has implications for the nature of the consultation that takes place.

In Australia, consultation is mediated by Human Research Ethics Committees HRECs (referred to as Institutional Review Boards in the United States) but is less prescribed than in the Canadian model. In New Zealand, the nature of the consultation in health research is also mediated by ethics committees (called Health and Disability Ethics Committees — HDECs) but the details about what counts as consultation and engagement is less detailed than in Australia.

As noted earlier, mandatory consultation in New Zealand has been reinforced by a set of guidelines produced by the Health Research Council of New Zealand entitled *Te Ara Tika* [Hudson *et al.* 2010]. These guidelines have proved useful in that they ambitiously address what constitutes best practice consultation while supporting the view that *all* health and tertiary research in New Zealand should go through consultation. By way of explanation, the absence of perceived social equipoise and prescriptive guidelines about how to engage with Indigenous concerns and

---

[2]Australian documents do not use the term Indigenous.

contexts has understandably promoted a much more cautious approach to recommended ethical oversight.

If the *Te Ara Tika* guidelines were to match the Canadian model they would need modification by making a case for New Zealand researchers to do less formal consultation but, when they do get involved in this, they are required to do it more thoroughly than is currently practiced. This change would entail all New Zealand researchers to address seriously, 'on paper', how their research will impact Māori, the description of which will be reviewed by an ethics committee. The implication is that only research that is at least Māori centered would require the researcher to actively consult with Māori, with clearer guidelines required to assist in the determination of the form that that consultation will take.

Te Ara Tika makes distinctions between mainstream research, Māori centered research and kaupapa Māori research (KMR). For the purposes of this discussion, KMR is excluded because first, its equivalent does not feature in any guidelines in Australia and, second, in New Zealand, by definition, Māori research with Māori communities by Māori will have culturally articulated consultation guidelines [see Smith 2018]. The focus of this conversation instead centers on the equation generating the greatest tension, namely how should non-Indigenous researchers deal with Indigenous contexts if they are interested in preserving cultural dignity and scientific credibility. In particular, a key question is whether consultation is generally restricted to matters of culture and the appropriate access to information and data or whether it should have wider functions and foci. In New Zealand and Australia, this type of re-negotiation of the research question is less prevalent with the focus appearing to remain centered on issues of culture and access. Here, the differences in the characteristics of what passes for Australian and Canadian Indigenous consultation compared to the 'pan research' format of consultation required in New Zealand is worthy of comment.

The social and economic conditions experienced by first peoples in Australia, Canada and New Zealand (briefly referred to in the Introduction) share several characteristics resulting from the colonization by settler peoples from Europe in the eighteenth and nineteenth centuries, with Indigenous scholars identifying the socio-political damage that emanates from arrangements that confirms and legitimizes situations of unequal socio-political capital [Rigney 1999; Ball & Janyst 2008; Smith 1999]. The impacts of this contact have been long lived with Indigenous peoples across the three nations exhibiting statistics indicating for outcomes for first peoples as reflected in social and economic disparity across a wide range of indicators. Data on incarceration, health status, mental health and life expectancy are remarkably similar in the three countries in terms of the degree of disparity between Indigenous and 'mainstream' populations. Of relevance to this dialogue though is the position that the nature of the research process has been central to and indicative of the continuation of injustice. As Linda Tuhiwai Smith [1999, p. 42] notes:

> Western research brings to bear on any study of Indigenous peoples, a cultural orientation, a set of values, a different conceptualization of such things as time, space, and subjectivity, different and competing theories of knowledge, highly specialized forms of language, and structures of power.

Variability in Emphasis in Approaches to Indigenous Consultation

Canada

The Canadian Tri-council Ethics Policy Statement essentially challenges researchers to: (1) become informed about, and to respect, the relevant customs and codes of research practice that apply in the particular community or communities affected by their research; (2) consider, where appropriate, to apply a collaborative and participatory research approach; (3) negotiate, where appropriate, a research agreement between the researcher and the community; and (4) resolve any discrepancies between institutional ethical policies and community customs and codes of research, by either adapting conventional practice or negotiating a resolution [Tri-Council Policy Statement 2018]. Unfortunately, these critical dimensions are not ones that are consistently or effectively dealt with, either as a part of research development, or as a primary function of ethics oversight in the New Zealand context.

The Canadian guidelines also prescribe positions on the value of local oversight, building research capacity in communities, and the importance of providing opportunities for Indigenous populations to shape the research questions. There remain tensions, particularly around the extent to which local communities should be empowered and have their wishes take precedence over those of the researcher. This can be seen when the guidelines stop short of giving communities the ability to block the publications of findings, preferring instead to view community input only in terms of 'contextualizing the findings'.

However, the guidelines do chart the end-points of the authority of community structures (as they pertain to research) by requiring that the researcher make allowances for the needs of individuals and subgroups that may not have a voice in the formal leadership, so as to extend community participation in the research project. In other words, the guidelines allow for patterns of community engagement that do not necessarily align with the more formal authority structures that are likely to play a key role in early interaction between researchers and the community. With regard to the distribution of harms and benefits, the Canadian guidelines do, however, give researchers working outside of accepted community authority structures reason for pause when it requires research to be conducted with an awareness of, and minimization of, the possibility of harm to the communities and individuals where research is undertaken that is critical of community institutions, whether Indigenous or not. Thus, in terms of the idea of social equipoise, there is equity in the quest for the minimization of harm.

Australia

The Australian guidelines pertaining to Indigenous peoples was updated in 2018, produced by the NHMRC in the shape of a short 29-page document entitled *Ethical Conduct in Research with Aboriginal and Torres Strait Islander Peoples and Communities: Guidelines for Researchers and Stakeholders*. Carried over from the 2007 document (*Values and Ethics: Guidelines for Ethical Conduct in Aboriginal and Torres Strait Islander Health Research*) the revised statement retains the six core values that it proposes should lie at the heart of research involving Indigenous communities. There is no explicit framework for the researcher to follow other than that any negotiation will be reviewed by a HREC. The six core values are: spirit and integrity, reciprocity, respect, equality, survival and protection and responsibility. For each value a range of considerations are detailed. For example, when demonstrating reciprocity, participating communities, researchers and HRECs should consider how the proposed research demonstrates intent to contribute to the advancement of the health and well-being of participants and communities; whether the proposal links clearly to community, regional, jurisdictional or international Indigenous health priorities and/or responds to existing or emerging needs articulated by Aboriginal and Torres Strait Islander Peoples; the nature of benefits for participants or other Aboriginal and Torres Strait Islander communities, and whether there is evidence of clear and truthful discussions about the potential benefit of the research proposal prior to approval; whether the researcher has demonstrated willingness to modify research in accordance with participating community values and aspirations and; whether the proposed research will enhance the capacity of communities to draw benefit beyond the project, e.g. through the development of skills and knowledge or through broader social, economic or political strategies at local, jurisdictional, national or even international level. However, the key change noted against the new edition of these guidelines is the broadening of the definition of health research resulting in the greater scope of applicability such that just about all human research involving Aboriginal and Torres Island peoples will be covered by this NHMRC statement.

The words 'should consider' suggests that these guidelines contain rules that are not necessarily governed or sanctioned by the Aboriginal and Torres Strait Islander communities, with *recommended* behaviors still to be mediated by the HRECs. The one overriding goal claimed in these guidelines is the elimination of 'difference blindness' which takes cognizance of the tendency for past research to trivialize the values and principles associated with Indigenous societies [see Smith & Tolich 2014 around the New Zealand experience]. Indeed, the word difference is mentioned sixteen times in the document with the sentiment summed up in the statement:

> Respectful research relationships acknowledge and affirm the right of people to have different values, norms and aspirations. Those involved in research should recognize and minimize the effect of difference blindness through all stages of the research process.

In Australia the focus of the guidelines is not so much concerned with the 'ownership' of research but rather on the recognition of a community's distinctiveness and what this means for the relationship between the researcher and the researched, whether this interaction centers on communities or individuals.

New Zealand

In comparison the situation in New Zealand is essentially straightforward in that all (health) research must involve a form of consultation even though the nature of that consultation is not clearly spelt out. By contrast, in both Australia and Canada, the types of research environment and characteristics where consultation is required is limited, with definitions and evaluations surrounding these limitations overseen by ethics committees (HRECs). New Zealand is thus unique with its blanket requirements that, on the surface, look as if it is set up to give optimal protection to Māori interests thus comfortably achieving a state of social equipoise. Unfortunately, the reality is somewhat different. As Hudson and Russell [2009, p. 61] point out, "while Māori values are acknowledged they are not given equal weight in ethical deliberations".

In New Zealand, it has been noted that guidelines are underpinned by the Treaty of Waitangi. The Treaty of Waitangi's three principles of protection, partnership, and participation are enshrined into ethical considerations. In some respects, the Treaty of Waitangi is New Zealand's original ethics statement. Thus, since the incorporation of Health and Disability ethics committees in 1988, both health and tertiary ethics committees have absorbed these principles into their ethics review processes when considering all research involving human subjects. There is some irony here as this has resulted in both a solution and a source of tension.

The opinion held by many Māori researchers in the health research sector, including those who sit as Māori members on ethics committees, is that consultation with Māori should occur except in very specifically defined circumstances [Smith 2011].

However, the view that instances of non-consultation should be seldom seen is one that generates a range of reactions within the health research community at large. As a response, criticism of the requirement to consult is generally couched in terms of the added cost and time arising from the difficulty of finding suitable parties with whom to consult and then having to wait for responses that are said to be often slow in coming. The extent of the general relevance of this process is also questioned. Solutions are then linked to the suggestion that consultation should only be required as a part of the ethics approval process when the research focuses specifically on Māori.

This perspective of course is a long way from the attitude that exceptions to Māori consultation should be rare since, as has been said, in terms of the Treaty of Waitangi, all health research is of relevance to Māori. Whilst approaches observed in applications to New Zealand Health and Disability Ethics Committees (HDECs) lie somewhere between these two stances, the annual reports from these committees as noted contain expressions of concern about the way researchers deal with Māori consultation, and cultural matters generally, which suggests that HDECs tend to favor a more broadly defined consultation requirement. Comments such as "Some researchers pay no more than lip service to the cultural requirements" [Ministry of Health 2008, p. 17] confirm a level of dissatisfaction with the status quo.

Of course, documents like the superseded 2006 *Operational Standard for Ethics Committees*, and other guidelines provided by the Health Research Council of New Zealand and the National Ethics Advisory Committee, highlight the importance of consultation while also underlining the principle of partnership with Māori. However, as a backdrop, views on consultation also bring to the surface attitudes about the function of health research and the purpose of ethics committees that are implicitly linked to ideas about the nature of the health issues faced by Māori whether viewed from a biomedical or sociological perspective and, by implication, the rights of Māori in New Zealand society and the persistent lack of equity in the distribution of harms and benefits.

Certainly, by way of reiteration, it is undeniable that there exists wide-ranging health disparity between Māori and non-Māori (with Māori faring worse) that is evidenced across a myriad of health indicators. For example, the observed differences in life expectancy of approximately eight years, the mortality rates that are nearly twice as high, and rates for amenable mortality that are almost three times as high indicate the existence of serious population and personal health issues. These statistics paint a challenging picture for the health sector even without broaching matters to do with condition specific differences and the observed widening of inequalities between regions [Pearce *et al.* 2008]. Furthermore, the persistent nature of this disparity makes it surprising that health inequality is not in the end considered a core driver of local health research that is explicitly reflected in both research funding priorities and ethical assessment criteria. The requirement that health researchers engage meaningfully with Māori across a whole range of research projects appears to be justified simply on the basis of the comparative state of Māori health.

Given this pervasive dimension, the absence of an appropriate level of engagement increases the likelihood that Māori will experience greater risk and receive fewer benefits from involvement in health research. At the very least then, any process that helps promote an equitable distribution of risks and benefits ('social equipoise') seen across different categories of participant as opposed to different interventions, would seem to be a not unreasonable goal for a nation's research community and systems of ethical oversight to be supporting. Considering this, and in the interest of social justice, the implication is that research projects need to have built into their design the capacity and opportunity to engage more fully and *purposefully* with Māori. The critical factor in this has to do with the integrity of that engagement defined in terms of both its nature and its intent.

To assist in this, *Te Ara Tika* defines potential behaviors in terms of an associated three-part hierarchy of ethical practice: minimum standard, good and best practice. These three levels are deemed appropriate according to whether the research is seen as mainstream, Māori centered or kaupapa Māori research, with the last category viewed as being an ideal given that it is defined as research that is designed and executed in accordance with Māori constructs and cultural processes. Minimum standards clearly expect a researcher to conduct a self-reflective exercise on how the project will impact Māori. Good practice in Māori centred research has the ethics committee review how the researcher has consulted with Māori in terms of the shape, the time scale and the extent of any interaction. This is different from Australia where mainstream research is exempt from consultative requirements.

Of course, kaupapa Māori research consultation given its definition demands little guidance on how to consult with Māori! This innovative paradigm instead provides scaffolding for Māori consultation where the theoretical and methodological foundations of the research are held together by purely Indigenous constructs. An important development in this space is recognition and support of this position in the revised National Ethical Standards for Health and Disability Research to be published by the National Ethics Advisory Committee (NEAC) that at the time of this writing is out for consultation.

Conclusion

In many ways, the colonization experiences of Australia, Canada and New Zealand are similar as is the intent of recent attempts to right past wrongs and yet, interestingly, the shape of the guidelines that define the nature of the relationship between research communities and Indigenous populations and communities remain different. The Australian mainstream research expectations are minimal but with guidelines calling for an end to 'difference blindness' when researching Indigenous communities. New Zealand requires that researchers should consult in essentially all circumstances which, it might be argued, has resulted in avoidance behaviour, poor practice and a form of systemic paralysis with regard to understanding and dealing with Indigenous environments.

Perhaps the ultimate solution is to be found in the concept of Kaupapa Māori research [see Smith 2018] which did not form part of this discussion because it is a paradigm that generally excludes non-Indigenous researchers, whereas the purpose of this chapter was to identify some of the sources of tension between non-Indigenous and Indigenous research and ethical constructs and environments. This is not to suggest that the Kaupapa Māori research context is always free of disagreement or debate but simply to verify that these elements were not the focus of the current conversations but is fully deserving of a detailed exploration.

However, in the New Zealand context, there has been a sense that maturation is being observed in relation to the way the tensions around Indigenous issues are being dealt with currently. For a start, the guidelines that circumscribe the behavior of the Health and Disability Ethics

Committees have just been re-drafted with an attempt to infuse the entire document with Māori constructs and definitions rather than have Māori matters confined to a separate appendix along with children and prisoners. Moreover, in this maturation vein, Māori scholars are simply taking the initiative and producing guidelines pertaining to advancing technologies such as genomic research [Beaton *et al.* 2017, Hudson *et al.* 2016] and data governance and health outcomes [Walker *et al.* 2017]. However, as is observed with much Indigenous activity in the global academic space, dialogue on ethical matters tends to mimic 'parallel play' where conversations tend to occur across nations between 'first peoples' rather than within nations between various socio-cultural entities. However, this aggregation of productive energy in global Indigenous environments may ultimately mark the best hope for managing ethical tensions in Indigenous research environments.

References

Ball, J. and Janyst, P. (2008). Enacting Research Ethics in Partnerships with Indigenous Communities in Canada: "Do it in a Good Way", *Journal of Empirical Research in Human Research Ethics* 3(2): pp. 33–51.

Beaton, A., Hudson, M., Milne, M., Port, R. V., Russell, K., Smith, B., *et al.* (2017). Engaging Māori in Biobanking and Genomic Research: A Model for Biobanks to Guide Culturally Informed Governance, Operational, and Community Engagement Activities, *Genetics and Medicine* 19(3): pp. 345–351.

Cunningham, C. and Stanley, F. (2003). Indigenous by Definition, Experience and World View, *British Medical Journal* 327(7412): pp. 403–404.

Hudson, M., Milne, M., Reynolds, P., Russell, K. and Smith, B. (2010). *Te Ara Tika: Guidelines for Māori Research Ethics: A Framework for Researchers and Ethics Committee Members.* Health Research Council of New Zealand, Auckland.

Hudson, M., Beaton, A., Milne, M., Port, W., Russell, K., Smith, B., Toki, V., Uerata, L. and Wilcox, P. (2016). *Te Mata Ira: Genomic Research with Māori.* University of Waikato, Hamilton.

Hudson, M. and Russell, K. (2009). The Treaty of Waitangi and Research Ethics in Aotearoa, *Journal of Bioethical Inquiry* 6(1): pp. 61–68.

Merges, R. P. (2011). *Justifying Intellectual Property* (Harvard University Press, Cambridge, Mass).

Ministry of Health (2008). *Northern Y Regional Ethics Committee Annual Report.* Available from <https://ethics.health.govt.nz/about-committees/archived-minutes-and-reports-pre-2012/northern-y-committee>.

Ministry of Health and University of Otago. (2006). Decades of Disparity III: Ethnic and Socioeconomic Inequalities in Mortality, New Zealand 1981–1999 (Ministry of Health, Wellington). Available from <http://www.rangahau.co.nz/assets/decades_disparity/disparities_report3.pdf>.

National Health and Medical Research Council (NHMRC). (2018). Ethical Conduct in Research with Aboriginal and Torres Strait Islander Peoples and Communities: Guidelines for Researchers and Stakeholders (National Health and Medical Research Council, Canberra).

Newbold, G. (2007). The Problem of Prisons: Corrections Reform in New Zealand Since 1840 (Dunmore Publishing, Wellington).

Pearce, J., Tisch, C. and Barnett, R. (2008). Have Geographical Inequalities in Cause-specific Mortality in New Zealand Increased During the Period 1980–2001, *NZ Medical Journal* 121: p. 1281.

Poata-Smith, E. (2013). Emergent Identities: The Changing Contours of Indigenous Identities in Aotearoa/New Zealand. In Harris, M., Nakat, M. and Carlson, B. (eds.) *The Politics of Identity: Emerging Indigeneity* (University of Technology, Sydney) pp. 56–59.

Reid, P. and Robson, B. (2006). The State of Màori Health. In Mulholland, M. (ed.) *State of the Màori Nation: Twenty-First Century Issues in Aotearoa* (Reed, Auckland).

Rigney, L. (1999). Internationalization of an Indigenous Anticolonial Cultural Critique of Research Methodologies: A Guide to Indigenous Research Methodology and its Principles, *Waikato SA Journal of Native American Studies Review* 14(2): pp. 109–121.

Rumball-Smith, J. (2009). Not in my Hospital? Ethnic Disparities in Quality of Hospital Care in New Zealand: A Narrative Review of the Evidence, *NZ Medical Journal* 122 (1297).

Smith, B. (2011). Màori Consultation, *HRC Ethics Notes, May,* Health Research Council of New Zealand, Auckland.

Smith, B. (2018). Màori Research: Notions, Limits and Possibilities. In Tolich, M. and Davidson, C. (eds.) *Social Science Research in New Zealand*, 4th edition (University of Auckland Press, Auckland).

Smith, B. and Tolich, M. (2014). A Cultural Turn: The Trivialization of Indigenous Ethics in New Zealand in the Post-2012 Health and Disability Ethics Committees, *MAI Journal* 3(2): pp. 255–267.

Smith, L. T. (1999). Decolonizing Methodologies: Research and Indigenous Peoples (Zed Books, London).

Theobald, M. (1993). Writing the Lives of Women Teachers: Problems and Possibilities, *Melbourne Studies in Education* 34(1): pp. 39–50.

Tolich, M. (2002). Pakeha Paralysis: Cultural Safety for those Researching the General Population of Aotearoa, *Social Policy, Journal of New Zealand* 19: pp. 164–178.

Tolich, M. and Smith, B. (2015). *The Politicisation of Ethics Review in New Zealand* (Dunmore Publishing, Wellington).

Tri-Council Policy Statement. (2018). *Ethical Conduct for Research Involving Humans*. Available from: <https://ethics.gc.ca/eng/documents/tcps2-2018-en-interactive-final.pdf>.

Walker, J., Lovett, R., Kukutai, T., Carmen, J. and Henry, D. (2017). Indigenous Health Data and the Path to Healing, *The Lancet*, 390(10107): pp. 2022–2023.

Wilkinson, R. and Pickett, K. (2009). The Spirit Level: Why More Equal Societies Almost Always do Better (Penguin Books, London).

CHAPTER NINE

'A Hungry Bushman is a Vulnerable Person': Considerations for Indigenous Bioethics

Julia Dammann,* Izak Van Zyl† & Danielle Pacia‡

Abstract

In March 2017, the San Council of South Africa, together with the South African San Institute (SASI), launched the San Code of Research Ethics. The main impetus for this was a genomic research project that was published in *Nature* in 2010 by Schuster *et al.* [p.943] While the authors obtained formal ethical approval and claimed to have video recorded consent, this research was deeply problematic to the San. This was primarily due to numerous conclusions, details and terminology used in the published paper that the San regarded as "private, pejorative, discriminatory and inappropriate" [Schroeder *et al.* 2016]. The problematic aspects were reported about under a European Commission funded project named TRUST [Schroeder *et al.* 2016]. It was under this project that the Code was subsequently developed by traditional leaders of the !Xun, Khwe and !Khomani groups of San, in partnership with civil society organisations, legal representatives and universities. This has been a

---

* Project Manager, South African San Institute, NGO, Kimberley, RSA.
† Associate Professor, Faculty of Informatics and Design, Cape Peninsula University of Technology, Cape Town, RSA.
‡ Candidate for Master of Bioethics, Harvard Medical School, Boston, Massachusetts, USA.

landmark development in the protection of Indigenous rights around research data.

In contrast to the above, Behrens [2013] finds that "[..] little academic work and few publications on bioethics reflect Indigenous African thought, philosophy or values" [p.33] With this chapter we would like to address the paucity of research on Indigenous bioethics, thereby substantiating and extending Behren's findings. Referring to the above-mentioned case and Code, we argue for a perspective that balances good science with the protection and preservation of indigeneity. We will support this argument by critically examining some of the aspects that continue to make the San vulnerable to exploitation and harmful research practice.

Introduction

The treatment accorded to Indigenous peoples by Europeans and Americans has been the antithesis of ethical [Wilson 2005, p.255].

The San communities in Southern Africa[1] have been of particular research interest to Western academia, not least in the fields of biomedicine, genomics and clinical research. In 2010, a prominent case study was published in *Nature*, detailing and comparing the genetic structures of Indigenous San peoples in Namibia and South Africa [see Schuster *et al.* 2010]. The study generated significant controversy in the way it portrayed members of the

---

[1] Some confusion still exists in the use of the terms San, Bushmen and Khoisan. The name San was given to the hunter-gatherer people by the neighboring Khoi people of the Cape. San means 'people different from ourselves' and the name was given to the San due to their stockless lifestyle. The name Khoi refers to the herding people of the Cape. Khoisan however is used by linguists to only refer to a certain group of click languages in Southern Africa, regardless of their hunting or herding cultural specifics. The San of today often refer to themselves by using their own groups names such as !Xun, Khwe or #Khomani (the different groups referred to in this paper). There are however a number of other hunter-gatherer groups living mainly in Namibia and Botswana. The term Bushmen, even though it stems from colonial times and is often considered derogatory, has regained popularity among different Bushmen groups. Clear categorization however is difficult as there are often no clear boundaries between the different cultural and linguistic groupings (Smith *et al.* 2000). In the anthropological literature, the different hunter-gatherer people are mostly referred to as San [Bahta 2014]. In this paper we will mostly refer to the !Xun, Khwe and #Khomani as San.

San in a derogatory and homogenous fashion [Chennells & Steenkamp 2018]. Moreover, the case raised important questions around informed consent: a requisite process in any ethical research inquiry involving human participants. Schuster's research team claimed to have adhered to all ethical regulations, on the basis of both video-recorded consent and institutional ethics approval in three countries. However, members of the San leadership were angered at their apparent exclusion, and at the lack of respect shown towards them.

This case, among many others, demonstrates that empirical and social scientific research has long been imbued with values of Western hegemony and imperialism. In the absence of critical, inclusive thought, Indigenous ways of knowing and being are marginalised or homogenised within Western paradigms of science [Smith 2012]. This holds particularly true in cases where Indigenous knowledge is being extracted for medicinal or other purposes, in the name of biomedical or clinical research [Stevenson & Murray 2016]. While ethical protocols govern exploitation and misappropriation of Indigenous knowledge systems, they too are generally rooted in Western modes of knowing [Coleman 2017; Israel 2017]. As a result, Indigenous communities become vulnerable in the face of Western-centred external research.

In a seminal book that explores the intersection between indigeneity and academic research, Smith [2012] lists 25 "projects" within Indigenous communities. These constitute different approaches communities are taking to make research more ethical, accessible and less exploitative. The 24th such approach, titled "discovering the beauty of our knowledge", concerns the integration of Western science and Indigenous knowledge in order to improve the circumstances of Indigenous communities. This approach is in line with the San Code of Research Ethics, published in 2017 as an output of a European Union project entitled TRUST: a pluralistic and multinational initiative aimed at "co-developing with vulnerable populations tools and mechanisms for the improvement of research governance structures" [EU TRUST 2019]. The Code has been developed in three workshops with a number of representatives of all three South African San communities over a period of about one year.[2] It comprises

---

[2] A detailed history of the San Code of Research Ethics can be found in Chennells and Schroeder [2019].

five pillars (values) as a means of ensuring informed consent, trust and inclusion during the research process.

Indeed, the San have historically been vulnerable to exploitation, as pointed out by Chennells [2009], who discusses the San world view based on hunting and gathering, collective trauma and the lack of access to land and resources as the primary factors of their vulnerability. The San Code is one means to combat invasive and non-inclusive research practice. In this article we aim to follow up on Chennells' factors of vulnerability and explore some of the aspects that continue to make the San vulnerable to exploitation and harmful research practice — in particular, resource scarcity, differences in notions of autonomy, cultural sensitivities, and language barriers. These combine to make the San communities in Southern Africa a site of struggle, in which various powerful actors vie for influence. In response to these facets, we first offer a critical overview of bioethical thought. Next, we unpack the Code and its potential for dismantling exploitation and unethical practice. We conclude the article with key insights and recommendations for future research.

Competing Values in Bioethics

> Any attempt to establish a secular bioethic must take seriously the pluralistic context of the postmodern world [Aulisio 1998, p.428].

Western categories of thought have dominated bioethical discourse in the sciences [Tangwa 2010]. Consequently, African and other heterogeneous perspectives, norms, and paradigms have not found a strong footing in modern frameworks of bioethics. The popular and original paradigm of bioethics, Beauchamp's and Childress's [2012] principlism, consists of four principles. Respect for autonomy, beneficence, non-maleficence and justice, that lie at the core of moral reasoning in health care. These principles "reflect a 'common morality' and can form the basis for all moral decision making" [Behrens 2013, p.34]. Among other concerns, the rigidity of the four principles has muffled other worthy values, like connectedness, solidarity and communal responsibility [Coughlin 2008]. The dominance of Western beliefs in bioethics leaves little room for nuance for diverse communities like the San to express the intricacies that come with their identity as both an Indigenous and African population.

Smith [2012] argues that research, dominated by Western thought, is an extension of the long history of exploitation and has argued that we should begin "decolonising" research methodology by recognising the various cultures associated with Indigenous communities. The general absence of diverse values in bioethical thought makes it challenging to create research paradigms that are considered pluralistically (or universally) ethical. Some values that are stressed in the Western perspective, like autonomy, are not always compatible with well-known maxims found across sub-Saharan Africa: "A person is a person through other persons" and "I am because we are" [see Mbiti 1990]. This idea of favouring the collective over the individual in ethical decisions is often referred to as "communitarianism" and is prevalent throughout sub-Saharan regions. Indeed, Coleman [2017] argues that the relational and communal nature of African morals is a "clear and dominant theme" throughout sub-Saharan Africa, a value relatively absent from Western discourse.

Bioethics is still often criticised for overemphasising individual autonomy and self-determination in non-Western contexts [Hunter & Leveridge 2011]. Western bioethics places a strong emphasis on gaining informed consent, namely, that respondents are fully informed of the study and its potential risks and thus agree to partake. The concept was created within the medical field, between doctors and patients, however, informed consent is now expected even outside of the medical field, especially when conducting research on human subjects. Although procedures are unique to each study, an ethical consent process normally conveys to participants their rights, as well as potential harms, benefits, risks, and study procedures. After being informed of rights, respondents should willingly participate in the study and researchers should attempt to avoid coercion as much as possible [Schroeder 2009]. However, the notion and act of gaining consent can become ambiguous when vulnerable and marginalised populations are included in a study. Because participation should be voluntary, the power dynamics associated with a subjugated group versus a dominant group may interfere with gaining ethical consent. It is requisite that parties consenting are ones that are "legitimate representatives", subjects that are able to understand and make decision effectively [Schroeder

2009, p.40]. For example, a five-year-old would probably not qualify as a legitimate representative because of their inability to properly understand potential repercussions of participating in the study. Likewise, populations who participate in research procedures that are not properly framed in terms of their cultural context can be considered an illegitimate representative, based on whether or not the researcher has taken the time to craft the project in a frame that is appropriate for the people being studied.

Although blatant expropriation has, mostly, come to end, Indigenous communities continue to be subjected to a more subtle form of exploitation through research. Schroeder [2009] explains that "[a]lthough blatant imperialist expropriation of lands and resources may have come to an end, large-scale economic development projects continue to disregard Indigenous values, interests and rights to participate" [Schroeder 2009, p.40]. The power dynamic (between a subjugated and dominant group) can infringe on one's autonomy by making it hard to willingly and freely choose to participate in the research, especially when the study promises certain types of benefits. In cases where food, water, and other essential resources are lacking in a community that is being approached for research, turning down an offer to participate in research that is coupled with benefits (that may supply the population with necessities) is increasingly implausible. In many academic circles (including bioethics), conducting research on prison populations is an example of how a vulnerable group's consent process must be carefully monitored because there is a higher potential of coercion [Moser et al. 2004]. In the context of the different San communities in Southern Africa, vulnerability may become a concern when those who are subject to resource scarcity are approached for research. Furthermore, many Indigenous populations still bear the effects of colonialism and other oppressive regimes which amplifies the risk of vulnerability and, in turn, amplifies the risk of coercion, creating circumstances that make proper consent difficult to achieve [Schroeder 2009].

When shifting across and between value systems and cultural contexts, acquiring consent becomes increasingly complex. Popular paradigms (like the Belmont Report and the Nuremberg Code) [Fischer 2006] are rooted in Western ideologies and stress the idea of autonomy throughout the consent process [Chattopadhyay & De Vries 2012]. Therefore, typically Central- and sub-Saharan values like communitarianism are often not properly accounted for during a research process affiliated with non-Western participants. The value of autonomy suggests that informed consent only needs to be gained from one participant, while in terms of communitarianism, this process can potentially extend to others — not just one individual. Apart from differing notions of autonomy, other heterogeneous aspects like language, social organisation, history and heritage affect the institution of empirical research with vulnerable, non-Western or Indigenous groups.

While there has been an overall improvement of informed consent procedures, Indigenous consent processes need to be strengthened. Remedying this power dynamic in order to create fair and equitable research hinges on pinpointing the aspects of populations that make them vulnerable [Schroeder 2009]. In the next section, we will attempt to highlight cultural values and aspects of the San that make it challenging to obtain proper consent and participate in research free of coercion and exploitation.

The San as a Vulnerable Research Grouping

An estimated 100,000 individuals living in southern Africa belong to a number of different San groups. They are linguistically and culturally diverse, with at least seven distinct languages and numerous sub-dialects. The social organisation of San groups differs due to the vast habitat of southern Africa with different climates, vegetation and game diversity. However, San groupings have in common a hunter-gatherer lifestyle [Chennells & Schroeder 2019]. Since approximately the beginning of the 18th century, many southern African San communities have been displaced due to colonisation and civil war [Smith *et al.* 2000]. Currently, the different San groups are generally impoverished and marginalised populations [Smith *et al.* 2000].

Three different San communities live in the Northern Cape of South Africa, namely the !Xun, Khwe and #Khomani. The !Xun and Khwe, now living on their own land called Platfontein close to Kimberley, were relocated from Angola and Namibia in 1990, after militarisation by the Portuguese International Police for Defence of the State (PIDE) in Angola and later, the recruitment to the South African Defence Force (SADF) in what is present-day Namibia [Van Wyk 2014]. The #Khomani San are originally from present-day South Africa. Their history was characterised by ongoing displacement through incoming pastoralists and later European settlers. The #Khomani became nearly extinct as they were incorporated into the lowest levels of the local economy, mostly as farm labourers. The #Khomani San's land rights were restored in 1999 when the new democratic government entitled them to six farms close to Askham in the Northern Cape. The #Khomani today live in Upington and in the southern Kalahari close to Askham [Pamo 2011].

For decades, the San's traditional lifestyle and knowledge, their history as well as their current situation has been of great academic interest. However, due to numerous complex reasons, research relationships have often been inequitable and even exploitative. As a result, the South African San published the San Code of Research Ethics on March 2, 2017. The Code was conceptualised to enable the San to tackle the complexity, inequity and potential exploitation that arise through empirical research. Aspects and values reflected in the San Code also inform the key insights in this chapter. Using the San as an example, we will explore the risks that occur when collaborative research is conducted between people from different cultural backgrounds. While we will not be able to portray the underlying complexity in this article, we will touch on topics that have made the San vulnerable to research. This is supported by a discussion of the San Code and its potential to minimise these risks and to dismantle exploitation and unethical practice.

Notions of Autonomy

Bahta [2014] argues that the present-day San communities in Platfontein have been corrupted and alienated from their culture, resulting in conflicting moral values and a struggle for identity. Furthermore, while traditional notions of identity and culture have become increasingly complex, they remain prevalent. This is also evident in the San Code of Research Ethics, where a strong sense of community is being expressed in the very first sentence:

> We require respect, not only for individuals but also for the community. We require respect for our culture, which also includes our history" (Value: Respect).

The San tend to view autonomy in terms of a broader system of communitarianism, with emphasis on collective values and equality. This different notion of autonomy has an impact on the research process, especially in the case of 'informed consent'. Because San groups emphasise community above individual autonomy, receiving consent from single individuals is regarded as disrespectful and holds the risk of doing harm to the community. The influence of cultural differences on the principle of autonomy (and on the consent process) is not a new phenomenon. It has been acknowledged and incorporated in the field of bioethics for some time; most prominently expressed in the example of a Jehovah's witness' refusal to receive a lifesaving blood transfusion due to religious reasons [Garraud 2014]. The landmark UNESCO Declaration on Bioethics and Human Rights (2005) contains an explicit article on respect for cultural diversity and pluralism. Despite this, cultural diversity may yet diminish in light of competing notions of autonomy and individual consent.

The potential harm that can be done if communities' value systems are not respected is illustrated in Chennells and Steenkamp's 2018 report of the Schuster genomic case. It provides an analysis of the consent process in respect of a genomic research project conducted in 2010. According to the report, consent was only obtained from the participating individuals while (a) genomic research speaks to collective issues, and (b) Indigenous communities have a different sense of individuality and individual rights. Not having obtained collective permission was perceived as disrespectful by the San. The authors conclude that respect for community is one of the most fundamental aspects of Indigenous research ethics [Chennells and Steenkamp 2018].

Through feedback given by the South African San Institute and participating San community members, the TRUST project identified "insensitivity to cultural differences" as one failure or challenge to ethical adherence in research with Indigenous communities. It was here found that a lack of written codes and model contracts for communities can be a gap in international research collaborations [Andanda *et al.* 2017]. It is therefore not a surprise that the San, when writing their Code, made sure that their value of community is articulated throughout:

> Research should be aligned to local needs and improve the lives of the San. This means that the research process must be carried out with care for all

involved, especially the San community. The caring part of research must extend to the families of those involved, as well as to the social and physical environment (Value: Care).

If not considered with sensitivity, the San's ideas and beliefs around autonomy make them potentially vulnerable to research. However, by ensuring that notions of community are embraced and respected, the San are able to shape the research process in terms of their values and thus reduce potential vulnerability.

Cultural Sensitivities

Some San groups such as those militarised in Angola were gradually exposed to Western knowledge and education systems from 1960 onwards [Van Wyk 2014]. In part due to the social and political disruptions over the past century, many San were driven to farm labour or militarised into the then SADF, preventing them from adapting to the national education system. In light of this, the different San groups have become vulnerable to highly specialised research encounters. With little knowledge about the concept of (modern) scientific research, the San are not equipped to protect their intellectual property. The Hoodia benefit sharing case, outlined below, illustrates the potential ethical dilemmas deriving from this imbalance.

Hoodia is a succulent plant Indigenous to Southern Africa. Research has shown that Hoodia and related species have been used by the San for centuries, mainly as a food and drink substitute due to its appetite suppressing attributes [Wynberg 2004, p.855]. Based on this, the Council for Scientific and Industrial Research (CSIR) implemented a project in 1963 to research edible wild plants of the region, with particular interest in the Hoodia plant. In 1995, the CSIR registered a patent application in South Africa for the use of the active appetite suppressing components of the Hoodia plant. License and royalty agreements were signed with Phytopharm and Pfizer, followed by international patent approvals. Until then, the San had neither been involved, nor acknowledged as traditional knowledge holders in this process.

In 2001, the non-government organisation (NGO), Biowatch South Africa, together with the international NGO Action Aid, triggered the discussion about the links between the patent and the traditional knowledge behind it. This development led to negotiations between the CSIR and institutions and individuals representing the San. Two years later, in 2003, a mutually acceptable benefit agreement was reached [Wynberg & Chennells 2009]. Wynberg [2004] writes that many challenges of the Hoodia case occurred because Western belief systems and managing approaches ran counter to those of the San, with the biggest ethical concern emerging before the commencement of negotiations. The sharing of knowledge has always been a matter of survival to the San: "Traditional knowledge of plants is viewed as a collective and the idea of "owning" life abhorrent" [Wynberg 2004, p.870]. Hence, the San shared their knowledge freely, with no possibility (or intention) of protecting it and themselves from exploitative practice. Schroeder [2009] points out in this regard that accessing traditional knowledge against the background of the legacy of colonialism might represent a continuation of colonialist exploitation with its inherent power imbalance.

The San refer to these cases in their San Code:

> We have encountered lack of justice and fairness in many instances in the past. These include theft of San traditional knowledge by researchers. At the same time, many companies in South Africa and globally are benefitting from our traditional knowledge in sales of Indigenous plant varieties without benefit sharing agreements, proving the need for further compliance measures to ensure fairness [Value: Justice & Fairness].

A paradigmatic imbalance however, is also very likely to exist on the researcher's side. With researchers from all disciplines being interested in working with the San, researchers may lack the necessary sensitisation towards the complex and dynamic cultural environment of San groupings. How this, if disregarded, can lead to possible harm, is revealed by the previously mentioned Schuster *et al.* [2010] genomic case. Problems occurred due to the researchers not being familiar with the cultural sensitivities of the participating San. This consequently manifested in the subsequent publication, in which the used language was perceived as derogatory by the San and conclusions irrelevant to genomic research were drawn [Chennells & Steenkamp 2018]. Furthermore, some of the researchers' interpretations were perceived by the San as "intimate, personal or pejorative information" [p.20].

These and other examples have raised distrust in research which reflects in the San Code of Research Ethics. The topics of educational difference and the consequent distrust occurs more than once in the Code and thus underlines its importance:

> We require an open and clear exchange between the researchers and our leaders. The language must be clear, not academic. Complex issues must be carefully and correctly described, not simply assuming the San cannot understand. (...) Open exchange should not patronize the San [Value: Honesty].
>
> We have encountered lack of care in many instances in the past. For instance, we were spoken down to, or confused with complicated scientific language, or treated as ignorant [Value: Care].

The San Code of Ethics anticipates the difference in belief and knowledge systems, as well as in educational background, and tailors the research process to ensure that the San remain protected.

Language

Another aspect related to the San's vulnerability in research is language. While the lingua franca in international research is English, the San's mother tongues are generally !Xun, Khwedam or Afrikaans. Furthermore, literacy among San communities is low [Pamo 2011]. Even in predominantly English-speaking countries, illiteracy creates challenges in garnering ethical consent [Muir & Lee 2009]. Generally, consent processes rely on participants reading, evaluating, and signing paperwork. But because of their history, much of the San population is illiterate. This creates a layered linguistic issue for outside researchers trying to conduct a fair consent process. Not only is there a marked difference in language, but also a shift in how the documentation may be recorded due to illiteracy. Essentially, the shift in procedure may cause researchers to miscommunicate the goals, risks, and benefits associated with the study — even if the forms were translated into the respective San language, there is no guarantee of 'informed' consent.

"The adequate comprehension of the disclosed information by the representatives" is one basic element of successful informed consent [Schroeder 2009, p.31]. Adding language and the literacy barrier to the difference in knowledge systems discussed in the previous section may impede the research and consent process. As Alaei *et al.* [2013, p.38] point out:

> Literacy and language are important factors for comprehension; previous studies have shown that both are educational barriers which may lead to poor comprehension or the lack of understanding consent information.

To deal with these challenges, the authors argue for their method of awareness to be a useful pattern for research with elderly and illiterate research participants. Following findings of studies that showed increased comprehension when information of the study is provided prior to the research, Alaei and colleagues included religious leaders in the informed consent process, allowed for time to decide, and gave the opportunity for the participants to observe the research process before giving consent. The researchers proclaim high numbers of participants and an ethical consent process.

The San Code of Research Ethics argues similarly, yet taking awareness a step further. The San demand to be involved throughout the entire research process. This includes engagement before the commencement of the research through to consultation in case of a subsequent publication. The San consider this involvement a matter of respect:

> Respectful researchers engage with us in advance of carrying out research [Value: Respect].

However, this is not just a matter of respect for the San. To ensure fair and appropriate participation, the San intend to implement this through their value of Process, where it says:

> The San research protocol that the San Council will manage is an important process that we have decided on, which will set out specific requirements through every step of the research process. This process starts with a research idea that is collectively designed, through to approval of the project, and subsequent publications [Value: Process].

Process is not only important to enable the participant to comprehend the research project, it can also help tackle the research dilemmas that Temple and Young [2004] point out. While it is common practice to use translators in research, it has been shown that even with translators, linguistic barriers impair the translation and interpretation process. Temple and Young [2004, p.164] draw attention to different translation dilemmas in research. Pointing out that some of the dilemmas are linked to the epistemological position of the researcher, they mention what has long been acknowledged by academics: "the importance of language in constructing as well as describing our social world". The translation process is never a neutral exercise [Temple & Young 2004, p.164]. Conceptual equivalence is impossible since "almost any utterance in any language carries with it a set of assumptions, feelings, and values that the speaker may or may not be aware of but that the field worker, as an outsider, usually is not" (Phillips 1960 in Temple & Young 2004, p.165].

Chennells and Steenkamp's [2018] report shows that the San have become entangled in these difficult situations. Very few San, of which most are in Namibia, manage to maintain a more traditional lifestyle living off the land and nature. The four San individuals that participated in the 2010 Schuster genomic project were of these San groups. According to the 2018 report, the San participants were illiterate and while the researchers claim to have received individual video consent, the authors question how the researchers communicated methodology, aims and objectives of a highly complex project.

Resource Scarcity

Over the past century, many San groupings were exposed to colonialism, border wars and apartheid that "led to modern post-foraging ways of living, such as village, reserve or settlement life, farming and herding, wage labour and other forms of employment" [Smith *et al.* 2000, p.82]. These aspects presented a series of new challenges: labour relationships were mostly paternalistic, unemployed San living on farms were considered illegal squatters; and disease, poverty and malnutrition proliferated [Smith *et al.* 2000, p.82]. As a consequence, the different San groups today generally belong to the poorest and most marginalised of populations as follows the social order instated by colonialism and apartheid.

The economic and political marginality of the San community presents a significant challenge and dilemma for outside researchers, and highlights the role and relation of incentives, payments, coercion and undue inducement in field research. Even minor financial incentives, for example, have been found to influence participants' decision towards partaking in the research [Largent *et al.* 2012]. Interestingly, the authors found that "a surprising 80% [of their sample] also judged that the offer of payment constitutes undue influence simply because it motivates someone to do something they otherwise would not" [Largent *et al.* 2012, p.6]. Conversely, Arnason and Van Niekerk [2009, p. 128] argue that inducement is unproblematic if the study is ethical, and go a step further: "International guidelines are mistaken to restrict inducements, to the detriment of those who would benefit the most: the poor and the vulnerable".

In a workshop conducted by the South African San Institute in October 2017, an elder San community member pointed out: "A hungry bushman is a vulnerable person". This statement opens up the discussion of a community's resource scarcity and general impoverished condition contributing to undue inducement or coercion. Furthermore, the voluntariness and autonomy of (collective) consent may be compromised because of this. Due to the San's participation in the TRUST project, undue inducement was not only identified within the medical research context, but also as a risk in empirical social scientific research with San communities. In this regard, we refer to the value of Respect in the Code, which explains that bribes and other advantages offered in order to secure approvals, is viewed as a lack of respect by the San [Singh & Schroeder 2017]. To tackle the dilemma inherent to payments, the San suggest "meaningful involvement" in their values of Justice and Fairness:

> It is important that the San be meaningfully involved in the proposed studies, which includes learning about the benefits that the participants and the community might expect. These might be largely non-monetary but include co-research opportunities, sharing of skills and research capacity, and roles for translators and research assistants, to give some examples [Value: Justice & Fairness].

Such an involvement leads to ethical reciprocity when participating in research. This in turn tends to result in a more equitable relationship between researcher and participant and consequently decreases the risk of potential inducement or coercion. Furthermore, under the value of Care, the San write about meaningful involvement:

> Research should be aligned to local needs and improve the lives of San. This means that the research process must be carried out with care for all involved, especially the San community [Value: Care].

This value is in line with recent discussions about the societal impact of research [Kirchherr 2018]. These discussions raise the concern that too much research is aimed at a particular scientific community rather than creating a positive impact for the non-academic community and/or its research participants. Furthermore, the TRUST project identified "Research priorities driven by Northern partners" as a risk factor that might potentially exploit local communities [Singh & Schroeder 2017, p.10]. Hence, some recent ethical guidelines — as cited by Singh and Schroeder — require that research be responsive to local priorities,[3] makes research mutually beneficial to the community being researched[4] and compensate appropriately through benefit sharing agreements in case of commercialisation of knowledge.[5]

The TRUST Global Code of Conduct emphasises the urgency to align research with local needs in its very first article under the value of Fairness:

> Local relevance of research is essential and should be determined in collaboration with local partners. Research that is not relevant in the location where it is undertaken imposes burdens without benefits" [Article 1, Global Code of Conduct for Research in Resource-poor Settings].

Consequently, if research is tailored towards the needs of the local community, it alleviates the harm associated with historical power dynamics and resource scarcity.

---

[3] E.g., Guideline 2, Council for International Organisations of Medical Sciences, 2016.

[4] E.g., Guideline 6.1.1 and 6.1.3, Indian Council of Medical Research, 2016.

[5] E.g., Guideline 6.7.8, Indian Council of Medical Research, 2016.

Concluding Thoughts and Recommendations

In this article we have attempted to emphasise the ongoing vulnerability of Indigenous communities to biomedical (and generally empirical) research. With particular reference to the San in South(ern) Africa, we described some of the foremost cultural, educational, linguistic and economic considerations that potentially conflict with Western modes of knowing. Engaging with 'resistless' communities in ways that do not accede to specific and sensitive conditions is arguably both unethical and irresponsible. The tension (and seeming incompatibility) between Western science and Indigeneity is not new. Historically, Indigenous peoples the world over have fought for their right to lands, waterways and oceans [Durie 2004]. These contests also transpire in intellectual and cultural sites, and are about the extent to which exclusive Indigenous knowledge can benefit greater society [Durie 2004]. With increasing modernisation, urbanisation and population growth, pressure is mounting on nation states to ensure an inhabitable future using all resources at their disposal. In this way, global development (not least in biomedicine) poses a continued threat to Indigenous groups; to their natural environments, traditional lifeways, values and beliefs.

For this reason, the scholarly community must continue to develop strategies to recover and preserve Indigenous knowledge [Simpson 2004]. This does not solely entail opposition to science as the only valid body of knowledge, or to the rejection of science in favour of Indigenous knowledge [Durie 2004]. Neither is it solely an indictment of the ethics behind the colonisation of Indigenous beliefs and practices [Wilson 2005]. What we propose, rather, is a deeper sensitivity toward and grasp of local dynamics — social, cultural, political, economic, environmental, and otherwise. This is not a new proposal, but one that is being neglected in the pursuit of scientific knowledge, as we have demonstrated throughout. The San Code of Research Ethics, conceptualised by representatives of the different San groupings, is one localised means of protecting Indigenous San groups from research exploitation. The Code reflects key values as entrenched in San cultural and belief systems as both a conduit and guideline for research activity. San communities, by and large, are by no means anti-research, but they do appeal that scientists "enter through the door and not the window" as so succinctly expressed by the late Andries Steenkamp, a respected San leader [San Code of Research Ethics 2017].

Finally, it may be useful to highlight the practicalities and future development of the Code. Chennells [2009] concluded that researchers must be aware of a community's vulnerability and comply with the requirements of engagement. While in the past the only route to fair involvement was through the researcher, the San have now taken charge and demand equal relationships through their Code. However, while the San Council in collaboration with the South African San Institute are the joint custodians of the Code, it is not yet punitively enforced (although this is expected to change). The primary ethical responsibility currently still lies with the researcher to read, internalise and effect the Code, in active consultation with San stakeholders. Thus, the Code becomes a set of deeply important conventions that supports the notion of egalitarian science. To remain relevant and significant, furthermore, the Code must develop over time, incorporating learnings from the experiences of others [see Hudson *et al.* 2010], and reach a deeper acknowledgement of local nuances that make the San vulnerable. In this way, the Code becomes (1) an innovative, participatory and localised instrument that protects indigeneity, and (2) a key example of 'localised ethics' in opposition to stringent ethics regulations in South Africa, based on universal Principalism imported from the Global North [Israel 2017].

Acknowledgements

This article is partly supported by the National Research Foundation (NRF) of South Africa and its Incentive Scheme for Rated Researchers. The article is also endorsed by the South African San Institute (www.sasi. org.za) as well as the South African San Council (+27 054 339 0327). We thank all stakeholders for their input and feedback.

References

Alaei, M., Pourshams, A., Altaha, N., Goglani, G. and Jafari, E. (2013). Obtaining Informed Consent in an Illiterate Population, *Middle East Journal of Digestive Diseases* 5(1): pp. 37–40.

Andanda, P., Wathuta, J. and Fenet, S. (2017). *Compliance Failures, a report for TRUST*. Available from: <http://trust-project.eu/deliverables-and-tools/>.

Arnason, G. and Van Niekerk, A. (2009). Undue Fear of Inducements in Research in Developing Countries, *Cambridge Quarterly of Healthcare Ethics* 18(2): pp. 122–129.

Aulisio, M. P. (1998). The Foundations of Bioethics: Contingency and Relevance, *The Journal of Medicine and Philosophy* 23(4): pp. 428–438.

Bahta, G. T. (2014). Cultural Conflict, Dilemmas and Disillusionment among the San Communities in Platfontein, *The Journal for Transdisciplinary Research in Southern Africa* 10(4): pp. 36–51.

Beauchamp T. L. and Childress J. F. (2012). *Principles of Biomedical Ethics,* 7th Edition. (Oxford University Press, Oxford).

Behrens, K. G. (2013). Towards an Indigenous African Bioethics, *South African Journal of Bioethics and Law* 6(1): pp. 32–35.

Chattopadhyay, S. and De Vries, R. (2012). Respect for Cultural Diversity in Bioethics is an Ethical Imperative, *Medicine, Health Care and Philosophy* 16(4): pp. 639–645.

Chennells, R. (2009). Vulnerability and Indigenous Communities: Are the San of South Africa a Vulnerable People? *Cambridge Quarterly of Healthcare Ethics* 18(2): pp. 147–154.

Chennells, R. and Schroeder, D. (2019). *The San Code of Research Ethics: Its Origins and History.* Available from: <http://www.globalcodeofconduct.org/wp-content/uploads/2019/02/SanCodeHistory.pdf>.

Chennells, R. and Steenkamp, A. (2018). International Genomics Research Involving the San People. In Schroeder, D., Cook, J. Hirsch, F., Fenet S. and Muthuswamy, V. (eds.) *Ethics Dumping — Case Studies from North South Research Collaborations* (Springer, Berlin) pp.15–22.

Coleman, A. M. E. (2017). What is "African Bioethics" as Used by Sub-Saharan African Authors: An Argumentative Literature Review of Articles on African Bioethics, *Open Journal of Philosophy* 7(1): pp. 31–47.

Coughlin, S. S. (2008). How Many Principles for Public Health Ethics? *The Open Public Health Journal* 1: pp. 8–16.

Durie, M. (2004). Understanding Health and Illness: Research at the Interface between Science and Indigenous Knowledge, *International Journal of Epidemiology* 33(5): pp. 1138–1143.

European Union (EU) TRUST Project. (2019). *Homepage.* Available from: <http://trust-project.eu/>.

Fischer, B. A. (2006). A Summary of Important Documents in the Field of Research Ethics, *Schizophrenia Bulletin* 32(1): pp. 69–80, DOI: 10.1093/schbul/sbj005.

Garraud, O. (2014). Jehova's Witnesses and Blood Transfusion Refusal: What Next? *Blood Transfusion* 12(Suppl 1): pp. 402–403.

Hudson, M., Milne, M., Reynolds, P., Russell, K. and Smith, B. (2010). *Te Ara Tika: Guidelines for Māori Research Ethics: A Framework for Researchers and Ethics Committee Members* (Health Research Council of New Zealand, Auckland).

Hunter, D. and Leveridge, J. (2011). The Concept of Community in Bioethics, *Public Health Ethics* 4(1): pp. 12–13.

Israel, M. (2017). Ethical Imperialism? Exporting Research Ethics to the Global South. In Iphofen, R. and Tolich, M. (eds.) *The Sage Handbook of Qualitative Research Ethics* (Sage, London) pp. 89–102.

Kirchherr, J. (2018). The Lean PhD: Radically Improve the Efficiency, Quality and Impact of Your Research (Red Globe Press, UK).

Largent, E. A., Grady, C., Miller, F. and Wertheimer, A. (2012). Money, Coercion, and Undue Inducement: Attitudes about Payments to Research Participants, *IRB Ethics and Human Research* 34(1): 1-8.

Mbiti, J. S. (1990). *African Religions and Philosophy* (Heinemann, Oxford).

Moser, D. J., Arndt, S., Kanz, J. E., Benjamin, M. L., Bayless, J. D., Reese, R. L., Paulsen, J. S. and Flaum, M. A. (2004). Coercion and Informed Consent in Research Involving Prisoners, *Comprehensive Psychiatry* 45(1): pp. 1–9.

Muir, K. W. and Lee, P. P. (2009). Literacy and Informed Consent: A Case for Literacy Screening in Glaucoma Research, *Archives of Ophthalmology* 127(5): pp. 698–699.

Pamo, B. (2011). San Language Development for Education in South Africa: The South African San Institute and the San Language Committees, *Diaspora, Indigenous, and Minority Education* 5(2): pp. 112–118.

San Code of Research Ethics. (2017). Available from: <http://www.globalcodeof-conduct.org/wp-content/uploads/2018/04/San-Code-of-RESEARCH-Ethics-Booklet_English.pdf>.

Schroeder, D. (2009). Informed Consent: From Medical Research to Traditional Knowledge. In Wynberg, R., Schroeder, D. and Chennells, R. (eds.) *Indigenous Peoples, Consent and Benefit Sharing, Lessons from the San-Hoodia Case* (Springer Science+Business Media B.V.) pp. 27–52.

Schroeder D., Cook, L. J., Fenet, S. and Hirsch F. (Eds) (2016). "Ethics Dumping" — Paradigmatic Case Studies, a report for TRUST, pp. 35–42. Available from: <http://trust-project.eu/deliverables-and- tools/>.

Schuster, S. C., Miller, W., Ratan, A., Tomsho, L. P., Giardine, B., Kasson, L. R., Harris, R. S., Petersen, D. C., Zhao, F., Qi, J. and Alkan, C. (2010). Complete Khoisan and Bantu Genomes from Southern Africa, *Nature* 463(7283): pp. 943–947.

Simpson, L. R. (2004). Anticolonial Strategies for the Recovery and Maintenance of Indigenous Knowledge, *American Indian Quarterly* 28(3/4): pp. 373–384.

Singh, M. and Schroeder, D. (2017). *Exploitation Risks and Research Ethics Guidelines, A Report for TRUST*. Available from: <http://trust-project.eu/deliverables-and-tools/>.

Smith, L. T. (2012). Decolonizing Methodologies: Research and indigenous Peoples (Zed Books Ltd, New York, US).

Smith, A., Malherbe, C., Guenther, M. and Berens, P. (2000). *The Bushmen of Southern Africa — A Foraging Society in Transition* (David Philip Publishers).

Stevenson, S. A. and Murray, S. J. (2016). Aboriginal Bioethics as Critical Bioethics: The Virtue of Narrative, *American Journal of Bioethics* 16(5): pp. 52–54.

Tangwa, G. B. (2010). *Elements of African Bioethics in a Western Frame* (African Books Collective).

Temple, B. and Young, A. (2004). Qualitative Research and Translation Dilemmas, *Qualitative Research* 4(2): pp. 161–178.

Van Wyk, A. S. (2014). The Militarisation of the Platfontein San (!Xun and Khwe): The Initial Years 1966–1974, *The Journal for Transdisciplinary Research in Southern Africa* 10(4): pp. 133–150.

Wilson, A. C. (2005). Reclaiming Our Humanity: Decolonization and the Recovery of Indigenous Knowledge. In French, P. A. and Short, J. A. (eds.) *War and Border crossings: Ethics When Cultures Clash* (Rowman & Littlefield, Lanham, NY) pp. 255–263.

Wynberg, R. (2004). Rhetoric, Realism and Benefit-Sharing: Use of Traditional Knowledge of Hoodia Species in the Development of an Appetite Suppressant, *The Journal of World Intellectual Property* 7(6): pp. 765–932.

Wynberg, R. and Chennells, R. (2009). Green Diamonds of the South: An Overview of the San-Hoodia case. In Wynberg, R. *et al.* (eds.) *Indigenous Peoples, Consent and Benefit Sharing: Lessons from the San-Hoodia Case* (Springer, Dordrecht) pp. 89–124.

CHAPTER TEN

Engaging Indigenous Pacific Island Communities

*Etivina Lovo**

Abstract

Principles of research bioethics are currently defined in the context of developed countries. Pacific Island nations' contextual meanings of such principles may be contrasting or complementing those that originated from developed countries that influenced the launch of the international guidelines of research involving humans. This chapter discusses these tensions.

Introduction: Inadequate Protection of Pacific Persons in Human Research

Research studies involving human participants have increased dramatically in many islands of the Pacific, but there has not been any corresponding increase in all areas of human research ethics [Denholm *et al.* 2017]. This is a cause for concern because the protection of human participants in research is inadequate in Pacific Island nations. A landmark case of this kind occurred in 2002, when Autogen, a Melbourne based

---

* Research Fellow, Fiji Institute of Pacific Health Research, College of Medicine, Nursing and Health Sciences, Fiji National University, Suva, Fiji.

Biotechnology Company, proposed a study to establish a database of genetic information on Tongan citizens. Many objected to the study, and one of the central issues concerned informed consent. The research ethics principles employed by Autogen in their proposal contrasted with those of the Tongan people. Autogen's perception of the principle of informed consent is that of individual rights to consent. This contrasted with the contextual meaning of informed consent in Tonga which is based upon interpersonal relationships within an extended family [Taufe'ulungaki *et al.* 2007]. This kind of collective consent will influence individual members of the extended family as to participating in research.

Another issue that was raised by this case is that of ethics review. In 2017, an article by Naka *et al.* [2017] found that:

> a missense variant, rs373863828-A (p. Arg457Gln), of the CREBRF gene (encoding CREB3 regulatory factor) was associated with an excessive increase in Body Mass Index in Samoans and the same gene was frequently observed in Tongans and was strongly associated with higher BMI. No significant association was detected in the three Melanesian and Micronesian populations living in Solomon Islands and Papua New Guinea [Naka *et al.* 2017, p. 847].

The article acknowledges the research participants for their "kind cooperation in providing blood samples and anthropometric data for testing. Informed consent was obtained from the patients" [Naka *et al.* 2017].

There was, however, no mention of an ethics application process nor an ethics review or approval from any of the participating countries' ethics committees. However, Tonga has had a National Health Ethics and Research Committee operational prior to 2014 [Tonga Ministry of Health 2014a].

The findings of the genetic study by Naka *et al.* have the potential to expose vulnerable Pacific peoples to many different forms of discrimination that may be caused by decisions made about employment, education opportunities or health insurance cover. Further, the study's lack of disclosure of an ethics application or an ethics approval from any government department of the Pacific countries involved in the research, indicate that the countries of study are open to exploitation in biomedical research by researchers of more powerful countries.

Ethical review structures have been developed in some Pacific Island nations. In the Republic of the Fiji, the government Ministry of Health and Medical services developed a National Health Research Guide in 2015 [Fiji Ministry of Health and Medical Services 2015] which is currently being revised because its original scope is insufficient. About a year after, Tonga developed a document called "Operational Guidelines for the National Health Ethics and Research Committee" by the Tonga Ministry of Health in 2014 based on the WHO's Operational Guidelines for Ethics Committees that Review Biomedical Research 2000 [Tonga Ministry of Health 2014b]. Apart from these documents there are very few publications about human research ethics from both countries and from other islands of the Pacific Region. Gopichandran [2017] suggests that human research ethics in low- and middle-income countries (LMICs) "lag behind". The reasons for the under-developed capacity in research ethics are due to lack of resources which includes the lack of expertise in bioethics, lack of a supporting system such as the legal framework, the lack of ethics in the educational systems [Gopichandran 2017].

More reasons for the under-developed status of bioethics in Pacific Islands are the lack of support for foundational activities, ethics, bioethics or clinical ethics education, governance of research bioethics reviews and career development of bioethicists [Gopichandran 2017]. What flows from this is a lack of capacity in Pacific Islands in bioethics. There is also lack of capacity to adopt the internationally accepted guidelines of human research ethics to include important Indigenous principles and practices.

Problems Experienced by Pacific Island Researchers

Meo-Sewabu [2014], a Pacific Indigenous researcher, exemplifies problems researchers face because of differences in understanding of human research ethics. She employed methods of research required in Fijian settings to uphold local Indigenous human research ethics principles but was not understood by academics and ethics committee in a university setting. Academic institutions currently restrict Pacific Island researchers to the use of research methods accepted internationally and do not include Indigenous methods based on Indigenous Pacific Island principles.

The environment in which human research ethics exist in the Pacific Islands has a knowledge framework structured within a paradigm of general truths and principles. This environment has not been examined to identify important elements that can be enhanced to strengthen human research and promote a high standard of ethics review. Therefore, there are two origins of human research ethics principles that guide research activities in Fiji and Tonga. Firstly, principles that are internationally accepted and documented in international guidelines and secondly, underlying Indigenous principles that guide the practices of peoples of Pacific origins. The sources of Indigenous principles are derived from local belief systems, culture, religious teaching and social structures. It is important to understand how these principles are utilized to inform human research ethics activities in country.

Colonialism Impacted the Pacific People's Understanding of Research Ethics

What does ethics mean in research in Pacific Island settings? In defining research ethics, culturally influenced ways of thinking are significant which makes us question how guidelines can be culturally accepted. For example, take the practice involved in the process of voluntary informed consent by research participants. The answer that I find in internationally accepted documents is that we begin the process by verbally asking the person for their permission to participate in a research and this person agrees and says "yes" and that is accepted and documented in a "consent form", thus ending the process. As mentioned earlier in the chapter, in Pacific settings the meaning of voluntary informed consent of the person to be involved in research is to go through a protocol involving various tiers in the family structure. The first step is to ask permission from the head of the extended family to conduct an extended family meeting to discuss the issue. The second step is to ask permission of the whole extended family as a group in a meeting. Thirdly, the permission of the head of individual nuclear family is sought and finally the permission of the individual. Thus, the voluntary informed process in Pacific setting is that of extended family relations and rather than individuals. This is an illustration of the distinctly different understanding of voluntary informed consent in Western and Pacific paradigms.

The example of the differences in interpretations of 'voluntary informed consent', first by internationally accepted documents and secondly by those of the Pacific person's understanding, could be directly related to Konai Helu Thaman's description of *"struggles"*.

> ...the struggles that our people have been going through over their years of education from basic education upwards — to first learn the dominant study paradigm and worldviews of western people who lived in other places at other times. This is "colonialism" on our Pacific nations' systems and their people's ways of knowing [Thaman 2003].

At a conference titled "Decolonizing Pacific Studies" in the University of Hawaii in 2003, Thaman stated that:

> ... much have been written about colonialism but little attention has been focused on its impact on people's minds, particularly on their ways of knowing, their views of who and what they are, and what they consider worthwhile to teach and to learn ...... [there is a] need to look critically at colonialism and how it has shaped our economies, our social structures and even our minds. [Pacific education is] an introduction to worthwhile learning, [therefore] if education is to be considered worthwhile for Pacific peoples, education should include "knowledge, skills, and values relating to the Pacific region and only then, it is worthy of being taught and of being learnt in Pacific Education [Thaman 2003].

Colonialism impacted the Pacific people's understanding of research ethics principles. Pacific Islands researchers face difficulties in designing a research method that satisfy the protocol of Pacific Islands settings but they are not understood by academics and by research programs in a university setting in countries outside of the Pacific region, or the research proposal being submitted to an ethics committee for review. The problem is caused by the perception that the internationally accepted guidelines of research ethics founded on Western philosophies are the only credible research ethics principles, while those principles of the Pacific are not important, devalued and worthless. Seiuli Luama Sauni voiced similar concern about research methodologies. He stated that:

> the current dominant research methodological paradigms evident in applied educational research methods are based on Western values and do not take cognizance of alternate yet equally valid research models or frameworks [Sauni 2011].

Decolonization of research bioethics is about recognizing how Western philosophies have dominated Pacific paradigms of research and that Indigenous principles and practices of Pacific research are to be given value and importance as they form the core of Pacific people's lives that makes living worthwhile. In re-valuing Pacific research bioethics, I need to highlight principles of research involving persons in Pacific Island nations that I believe are not included in the Western models of Research Bioethics.

Firstly, I want to discuss *'Cultural Discernment'* by Meo Sewabu [2014], in order to illustrate the difficulties faced by a Pacific researcher based on colonialism. Cultural discernment is a process in which a community or a group of people collaborate to ensure that the research process is ethical within the cultural context. Cultural discernment comprises of a group of relatives of the researcher who are selected to be part of the research group. Their role is to make sure that all protocols of cultural expectations are observed properly while the research is being carried out in the Fijian community. Cultural discernment is a process that is required by the Fijian Vanua Research Framework [Nabobo-Baba 2008]. If a researcher who is researching in Fiji does not observe cultural discernment, it is considered disrespectful and the research will not receive the blessings of the Fijian community resulting in the researcher facing multiple problems in collecting data [Meo-Sewabu 2014].

Indigenous Pacific Islands' Research Ethics Principles Not Found in International Research Ethics Guidelines

Further to the variations of the prominent values of research ethics in comparison to Pacific research ethics principles, I want to discuss principles of research in Pacific nations that are not found amongst the prominent values of research ethics found in international guidelines of research involving human beings.

Out of the documented Pacific Indigenous research frameworks from various Pacific Island countries, I studied three, the Fijian Vanua Research [Nabobo-Baba 2008], The Samoan Ula [Sauni 2011] and the Tongan Kakala [Fua 2014] research frameworks.

In my brief comparative analysis of these three Pacific Indigenous research frameworks, I identified Indigenous principles that, in my view, are not included in the prominent internationally accepted principles of research involving human beings.

I will name each one using the Indigenous terms and then provide a brief English translation followed by a description of how it is applied in the Indigenous setting. I will also mention whether this Indigenous value is common or not to all three countries of Fiji, Samoa and Tonga.

Feagaiga (Covenant): The Samoa Ula Research Framework extends the value of love, respect and trust to include *Feagaiga*, translated into English as *Covenant*. The meaning of covenant includes the word "bond". In this context there is a long-term bond between researcher and research participants. Sauni [2011] contextualizes *Feagaiga*:

> *Feagaiga* is a sacred and honourable principle within a Samoan *aiga* (family) context. The Samoan saying, *"O le ioi mata o le tama lona tuafafine"*, translates as, a sister is the pupil of her brother's eye and therefore needs to be protected. According to Tupuola [2000], the term *Feagaiga* refers to a sacred obligation for males to honor, respect and protect their sisters. In research, men treat women researchers respectfully and vice versa. For example, whenever women researchers were invited to the men's homes, a woman researcher was given the privilege of conducting the prayer before the meal was served [Sauni 2011, p. 60].

The principle of *Feagaiga* emphasizes that there must be a respectful relationship in research similar to the emotional respect between brother and sister. This respectful relationship bonds the researcher and the research participants just as a brother and sister are bonded through their bloodline. This respect is expected to be observed in every research involving Pacific peoples everywhere. The researcher establishes an emotional bond with the research participants and community like family.

Feagaiga is related to *feveitokai'aki* in Tongan and *veivakarokoroko-taki* in the Fijian Vanua Framework means that the respect is reciprocal, and the relationship will be lifelong.

The Fijian Vanua and the Tongan Kakala Researh Frameworks emphasize the importance of humility in a researcher. **Yalomalua** (Fijian Vanua) *and* **Loto fakatokilalo** (Tongan Kakala) *mean humility* in English. Researchers are to be humble at all times, by using respectful language, dress appropriately, give the research participants ownership of the research project.

Na vakavinavinaka means gift giving and receiving in the Vanua Research Framework and has the same meanings with **Meaalofa** in the Ula framework and **Fe'inasi'aki** in Tongan, where gifting is reciprocal between the researcher and research participants. Reciprocity of gifting involves sharing of resources including knowledge, skills and other tangible products.

Spirituality is found in the Vanua Research framework, which means **Lotu** and it is also defined as *spirituality that is non-religious,* that of the Spirit worlds, beliefs, knowledge systems, values and Gods. There is a distinction between permission and blessing granted to a researcher in Fiji. Permission is only half of the deal. A researcher also needs the blessings from the *Vanua* (Fijian land and people). Therefore, a local protocol **sevusevu** (the presentation of yaqona 'piper methystica' to request entry into the community) is to be conducted in order to receive the blessing of the Vanua. Spirituality in Vanua is also found in Ula. **Fa'aleagaga** in Ula means spirituality in connection to Christianity, traditional beliefs and spiritual ancestors. *Fa'aleagaga* is *mana* (supernatural power) for Samoan people because of the connections to their Christian faith, traditional beliefs and spiritual ancestors. Kakala framework discusses the spiritual belief in Christianity as **Fakalaumalie.**

Veiqaravi in Vanua means to serve or service to Vanua which is the same as **Tautua** (service) in Ula. *Tautua* (service) refers to the roles and responsibilities in traditional customs, which include the young serving the adults and their elders, that is, the way to authority is through service. **Faifatongia** *(service)* in Tongan means the same as in the Vanua and Ula.

Gagana (appropriate language) in Ula. Language is essential and fundamental for bridging understandings, translations of conversations and maintaining relationships within Samoan culture. Respectful languages are emphasized by the Vanua Research Framework, and Kakala, *lea faka'apa'apa* is a requirement of researchers or their representatives, to be fluent in the Fijian dialect and Tongan language.

Indigenous Pacific knowledge expressed by Kakala, Ula and Vanua is sacred *Tapu or taboo*. Indigenous knowledge is a blessing that belong only to certain family lines and is kept sacred amongst people of that family line. To ask for a share of Indigenous knowledge is asking the Indigenous peoples to break the *tapu* for the benefits of the general population and for future generations.

My Fatongia

This is my responsibility or my *fatongia* to my people and the people of the Pacific islands to promote the engagement of our identities, our physical beings, our *Vanua*, our *Lotu*, our Spirits and our voices in ethical biomedical research.

Therefore, it will be my challenge to seek and try to explain theories of the principles that inform our ethical approaches to the ethical conduct of human research to be conducted about our lives. Research bioethics principles unique to Pacific peoples are to be promoted so that research scientists worldwide are aware of their roles when conducting research involving Pacific peoples. A model framework is to be developed of a harmonized Indigenous and international approach to human research ethics so that we are not just the 'researched' but we partner, we belong, we own. I may not know exactly how to take up these challenges, but I know that I will seek out those with the sacred knowledge in our islands and ask them to share it with the world.

References

Denholm, J., Bissell, K., Viney, K., Durand, A., Cash, H., Roseveare, C., ... Biribo, S. (2017). Research Ethics Committees in the Pacific Islands: Gaps and Opportunities for Health Sector Strengthening, *Public Health Action* 7(1): pp. 6–9.

Fiji Ministry of Health and Medical Services. (2015). *Fiji National Health Research Guide.*

Fua, S. U. J. (2014). Kakala Research Framework: A Garland in Celebration of a Decade of Rethinking Education (USP Press).

Gopichandran, V. (2017). Development of Capacity for Research Ethics Review in Low-and Middle-Income Countries: Need for a Systems Approach, *Public Health Action* 7(1): p. 1.

Meo-Sewabu, L. (2014). Cultural Discernment as an Ethics Framework: An Indigenous Fijian Approach, *Asia Pacific Viewpoint* 55(3): pp. 345–354.

Nabobo-Baba, U. (2008). Decolonising Framings in Pacific Research: Indigenous Fijian Vanua Research Framework as an Organic Response, *Alternative* 4(2): pp. 140–154.

Naka, I., Furusawa, T., Kimura, R., Natsuhara, K., Yamauchi, T., Nakazawa, M., Matsumura, Y. (2017). A Missense Variant, rs373863828-A (p. Arg457Gln), of CREBRF and Body Mass Index in Oceanic Populations, *Journal of Human Genetics* 62(9): p. 847.

Sauni, S. L. (2011). Samoan Research Methodology: The Ula-A New Paradigm, *Pacific-Asian Education Journal* 23(2): pp. 53–64.

Taufe'ulungaki, A., Johansson Fua, S., Manu, S. and Takapautolo, T. (2007). *Sustainable Livelihood and Education in the Pacific Project — Tonga Pilot Report* (Institute of Education, University of the South Pacific, Suva).

Thaman, K. H. (2003). Decolonizing Pacific Studies: Indigenous Perspectives, Knowledge, and Wisdom in Higher Education, *The Contemporary Pacific* 15(1): pp. 1–17, DOI:10.1353/cp.2003.003.

Tonga Ministry of Health. (2014a). National Health Ethics and Research Committee.

Tonga Ministry of Health. (2014b). Operational Guidelines for the National Health Ethics and Research Committee.

Tupuola, A.-M. (2000). *Raising Research Consciousness the Fa'asamoa Way.* University Institute of Cultural Policy Studies, Brisbane, Australia.

Index

Printed in the United States
by Baker & Taylor Publisher Services